# Community Paleoecology as a Geologic Tool:

## The Chinese Ashgillian-Eifelian (latest Ordovician through early Middle Devonian) as an example

Wang Yu, A. J. Boucot, Rong Jia-yu, Yang Xue-chang

SPECIAL PAPER

211

# Community Paleoecology as a Geologic Tool:

## The Chinese Ashgillian-Eifelian
## (latest Ordovician through early Middle Devonian)
## as an example

**Wang Yu**
(Deceased)
Nanjing Institute of Geology and Palaeontology
Academia Sinica
Nanjing, People's Republic of China

**A. J. Boucot**
Department of Geology
Oregon State University
Corvallis, Oregon 97331

**Rong Jia-yu**
**Yang Xue-chang**
Nanjing Institute of Geology and Palaeontology
Academia Sinica
Nanjing, People's Republic of China

# SPECIAL PAPER
# 211

Published by The Geological Society of America, Inc.
3300 Penrose Place, P.O. Box 9140, Boulder, Colorado 80301

Printed in U.S.A.

GSA Books Science Editor Campbell Craddock

**Library of Congress Cataloging-in-Publication Data**

Community paleoecology as a geologic tool.

  (Special paper ; 211)
  Bibliography: p.
  1. Paleoecology—China. 2. Animal communities—
China. 3. Paleontology—Ordovician. 4. Paleontology—
Silurian. 5. Paleontology—Devonian. 6. Paleontology—
China. I. Wang, Yu. II. Series: Special paper
(Geological Society of America) ; 211.
QE720.C66    1987       560'.45'0951       87-7549
ISBN 0-8137-2211-X

# *Contents*

# *Acknowledgments*

This study is part of an overall collaboration by Boucot and several members of the Nanjing Institute on varied problems of Silurian-Devonian community paleoecology, biogeography, and Silurian correlation. These problems are intimately involved with each other, and with the basic taxonomic research in progress by all of us.

We are grateful to the Academia Sinica for full support. Boucot thanks the U.S. National Academy of Sciences' Committee on Scholarly Communication with the People's Republic of China for support of his activities, as well as the Nanjing Institute of Geology and Palaeontology, Academia Sinica, and its personnel for the many courtesies received during his work with scientists of that institution.

We thank the following individuals for their generosity in providing a wealth of unpublished information, advice, and field assistance necessary for this study: Xu Han-kui and Zhang Ming, Nanjing Institute of Geology and Palaeontology; Fu Li-pu and Zhang Yan, Xi'an Institute of Geology and Mineral Resources, Xi'an, Shaanxi; Yan Guo-shun, Henan Institute of Geological Science, Zhengzhou, Henan; Zhang Zi-xin, Regional Geological Survey Team of Xinjiang, Qitai, Xinjiang; Fang Run-sen, Yunnan Institute of Geological Science, Kunming, Yunnan; Su Yang-zheng, Shenyang Institute of Geology and Mineral Resources, Liaoning; Kuang Guo-dun, Guangxi Bureau of Geology, Nanning, Guangxi; Chen Yuan-ren, Geological College of Chengdu, Chengdu, Sichuan. We are indebted to Peter Sheehan, Milwaukee Public Museum, Milwaukee, Wisconsin; Derek Ager, University of Wales, Swansea; and Kenneth R. Walker, University of Tennessee, Knoxville, for critically reviewing earlier versions of the manuscript. We are particularly indebted to J. Thomas Dutro, Jr., U.S. Geological Survey, Washington, D.C., for his painstaking reviews of several versions of the manuscript.

Geological Society of America
Special Paper 211
1987

# Community Paleoecology as a geologic tool: The Chinese Ashgillian-Eifelian (latest Ordovician through early Middle Devonian) as an example

## ABSTRACT

For the first time the faunal data for the Chinese Ashgillian through Eifelian interval are presented in a community, a community group, and community evolution context. These Chinese data are compared and contrasted with similar data for the same time intervals available from other parts of the world.

A brief review of the techniques of community analysis is provided. Emphasis is placed on the nature and preparation of community frameworks, and on the use of the community group concept in classifying communities into a form more useful for purposes of basin analysis.

We conclude that community evolution, under appropriate conditions of biogeographic isolation from many other parts of the contemporary marine world, proceeded in China along lines similar to those recognized elsewhere in the world. Our data are cast into a basin analysis framework, time interval by time interval, through the use of 11 community framework diagrams.

More than 60 community units are defined and described, most of which are assigned to a community group.

Most of the communities described here from the Chinese Ashgillian through Eifelian interval represent relatively near-shore, shallow-water, photic zone equivalent, Benthic Assemblage 2 and 3 communities. Notable too is the relatively small number of communities, representing an equivalent or even smaller number of community groups, for each of the time intervals selected for community analysis. This indicates the importance of analyzing a very large part of the world when doing such work, and also the need to compare conclusions derived from one area with work done in another area for internal consistency.

The high level of endemism characteristic of most of the Chinese Ashgillian through Eifelian ensures that most of the communities from this part of the Paleozoic will be unique, rather than shared with other, previously analyzed parts of the world such as North America and Europe.

## INTRODUCTION

Although fossils traditionally have been used chiefly for determining the relative ages of stratified rocks, they also have provided some help to the geologist concerned with environments of deposition and basal analysis. From time interval to time interval, and from place to place, terms such as reef, graptolitic facies, nearshore fauna, deep-water fauna, oyster bank, biohermal facies, and biostrome have proved to be useful concepts. There has never been a regular, routine effort, however, to cast the data of paleontology—available for each region and time interval—into an environmental classification applicable to basin analysis and environmental interpretation—a puzzling omission. For some time geologists have employed a variety of sedimentologic and geophysical parameters for basin analysis. Paleontologic biofacies data can be sorted out for basin analysis purposes, just as easily as grain size, sorting, and texture. In principle, combinations of physical and paleontologic parameters should provide significantly higher levels of resolution in basin analysis than can be obtained with either alone. Herein, we present an example showing how routine paleontologic data, of varying qualities and obtained by different investigators at different times and places, may be combined and synthesized for potential use in basin analysis. Although this example is taken from the Ashgillian through Eifelian interval in China, the principles and analytic approach may be used anywhere in any part of the Phanerozoic.

The first step is to prepare a community analysis. Keep in mind that paleontologists concerned with correlation and age dating by means of fossils emphasize the *similarities* among faunas and floras. Consequently, there is commonly a great deal of discussion about shared species and genera. In basin analysis, paleontologists reverse their position, emphasizing the *differences* among faunas and floras while simultaneously trying to unravel the environmental significance of these differences. The presence of faunal differences provides a clue to environmental differences. The paleontologist is aware that similar sedimentary rocks, occurring at different localities and representing the same time horizon, as well as occurring at different horizons, yield distinctive faunas that suggest environmental differences not obvious to the sedimentologist. For example, the very presence or absence of organisms can distinguish between the thixotropic-dilatant properties of sedimentary rocks in places where the parameters used by the sedimentologist are unable to make such distinctions. Fossils give the paleontologist information about the nature of the sediments that is not readily apparent from a study of standard physical parameters. The paleontologist sorts out the biotas into as many distinctive, recurring biofacies as possible. We prefer the term community to biofacies, because the latter has a possible ambiguity with biogeographic units. In addition, the term community has appropriate biologic, including ecologic and evolutionary, significance.

After the fossil biotas within a specific region and stratigraphic interval have been classified into community units, they should be checked against similar units known and described from similar time intervals elsewhere in the world. For example, in our Chinese work we built on knowledge concerning similar communities described from Europe, North America, and Australia. Closely related organisms, commonly those belonging to the same family, behave environmentally in very similar ways no matter what part of the world they inhabit. This behavioral conservatism of most organisms makes it possible to check, for internal consistency, environmental interpretations made in one region with those made in another region. For example, shallow-water, commonly nearshore, brackish water oyster banks occur from the Cretaceous to the Holocene and possess similar environmental significance wherever they occur. When seeming anomalies are uncovered, it is possible to assess critically the data leading to different environmental interpretations to see whether they require unique solutions. Most other, less familiar, organisms behave in a predictable manner, similar to the well-known oyster bank example.

A mere systematic tabulation of community data is insufficient for a paleontologic basin analysis program. The community data must be organized, horizon by horizon (a certain level of evolution is occurring that requires some concern for time-related faunal changes), into a community framework (Figs. 7-17 are examples using the Chinese data provided by our study). A community framework (see Boucot, 1981, p. 247–251; 1982, for discussion and examples) is a means of systematically organizing the many communities that were simultaneously present within a region during any one time interval.

What is a community framework? Why employ one? How may one construct one? Simple inspection of the "Alphabetical List of Communities," which comprises the second half of this paper, should convince even the most determined reader that trying to keep such a mass of faunal and lithologic data in mind for varied purposes is virtually impossible. Yet it is essential to be able to compare the environmental significance of the varied communities belonging to the same time interval—lateral biofacies relations—and also be well aware of the vertical similarities and differences through time that involve evolutionary changes within the same biofacies (termed a community group). It must also be noted which community groups are present or absent, time interval by time interval, within any one area.

The Community Framework (Fig. 1) is constructed as follows: Along the vertical ordinate, proximity to shoreline is laid out. This vertical ordinate is both a measure of nearness to shore and depth. In general, distance from shoreline usually but not always correlates with increasing depth, because there are platform margin shallows in some situations. The horizontal abscissa divides communities using both biologic and physical parameters such as species diversity (high, low, medium—with low diversity commonly indicating restrictive conditions), turbulent environments, quiet-water environments (indicated by laminated, nonbioturbated sediments), high salinity, low salinity, high oxygen, low oxygen, and sediment grain size (Boucot, 1981, 1984). It is along the abscissa that knowledge of the varied physical parameters commonly used in basin analysis can be displayed.

The community framework enables researchers to decide just what the environmental relations of a new community unit might be, not only by taking advantage of the physical and biologic data provided by the new community unit itself, but also by noting its lateral and vertical stratigraphic relations to previously studied community units whose environmental significance has been determined in other places or within the same region. Such a community framework allows arrangement of the varied communities consistent with stratigraphic trends indicating shallowing and deepening, increase or decrease of turbulence, proximity to shoreline, and the like. It also incorporates biotic data that, in some instances, may provide finer divisions than do the physical data alone. After the framework is constructed, it can be checked both against the community sequence present in different stratigraphic sections and against the communities present at the same horizon, laterally in various directions—shoreward or basinward.

The recognition of communities—regularly recurring taxic associations with relatively similar abundances of the taxa—and the preparation of community frameworks are followed by sorting out similar time-sequence communities into evolving community groups. A community group may be defined as a regularly recurring association of genera—some common, others

Figure 1. Outline diagram of a community framework, with the ordinate devoted to characters that vary with proximity to shoreline, and the abscissa to characters that vary parallel to shoreline. Any number of parameters may be employed, depending on the nature of the available data. The empty boxes are used for individual communities; their boundaries tend to reflect the statistical data available for each individual analysis. There is, of course, a tendency for community upper and lower limits to cluster about such critical points as mean low tide and the base of the photic zone, but there is no reason to suspect that most boundaries should have the same shoreline proximity value. (Modified from Boucot, A. J., 1981, Fig. 207, p. 250, Principles of Benthic Marine Paleoecology, New York, Academic Press.)

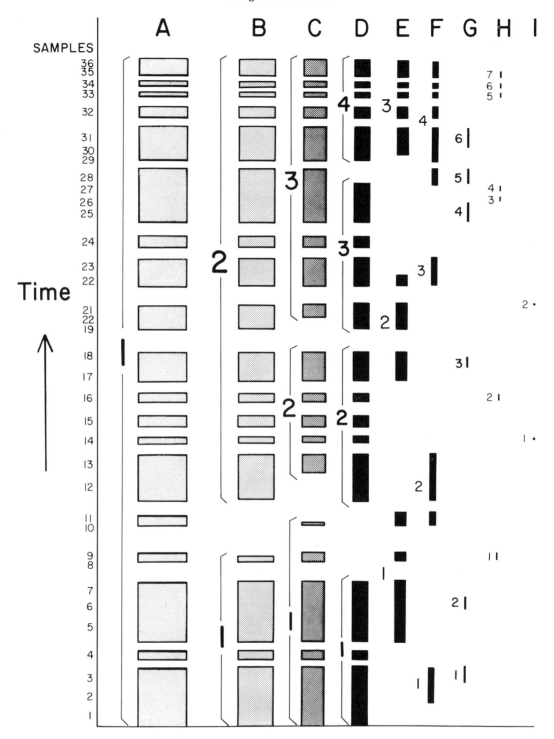

Figure 2. Diagrammatic outline of the evolutionary pattern characteristic of the genera present in a specific community group (capital letters indicate genera; numerals, the species). Note the inverse relationship between the rate of phyletic evolution and the abundance of each genus (indicated by line width). Gaps in the sampling are generalized, and can correspond to covered intervals, erosional intervals, nondeposition, or structural breaks. (From Boucot, A. J., 1983, Fig. 2, Journal of Paleontology, v. 57, p. 3.).

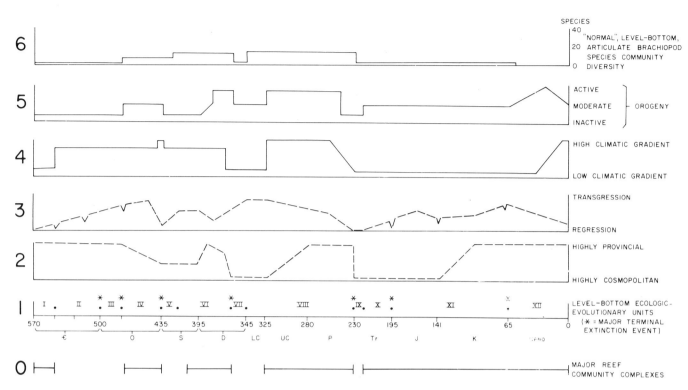

Figure 3. Diagram indicating generalized correlates of the basic, major biostratigraphic, ecologic-evolutionary units into which the Phanerozoic fossil record may be divided. (From Boucot, A. J., 1983, Fig. 3, Journal of Paleontology, v. 57, p. 8.)

rare, some intermediate. The less common genera within a community group tend to have rapidly evolving species (Fig. 2), whereas the reverse is true for the more common genera. Next, the community frameworks prepared for each horizon are checked to see if communities belonging to the same evolving community group occupy a consistent position within the community frameworks. If not, careful consideration must be given as to why the differences exist and whether they are truly necessary. In some instances, such a time sequence–community framework analysis will require revision of some of the environmental interpretations. In any event, using such an analysis the paleontologist can prepare an internally consistent community environmental interpretation for as many horizons as there are available data.

The common marine fossils of chief concern to most paleontologists and geologists occur in a limited number of ecologic-evolutionary units (Fig. 3) (Boucot, 1983). Phanerozoic level-bottom marine faunas may be divided into 12 major ecologic-evolutionary units, each characterized by a large number of community groups. Most community groups occurring in each ecologic-evolutionary unit do not recur in other ecologic-evolutionary units; this is merely a way of formally recognizing

that there are a limited number of Phanerozoic biofacies occurring through time. For example, oyster banks do not occur before the Cretaceous, pentameroid brachiopod biostromes are characteristic for part of the Silurian but not for the Ordovician, and so on. This ecologic-evolutionary unit characteristic allows the paleontologist, working within a specific time interval, to take advantage of all the data from varied parts of the world when using fossils in a basin analysis context. This worldwide ecologic sampling is analogous to the valuable biostratigraphic data also available for most fossil groups on a global basis.

Organisms from the base of the Cambrian to the present occur in irregular patches (Fig. 4), and no two parallel transects from shoreline seaward are ever precisely identical physically or paleontologically. Thus a methodology capable of dealing with this material in a systematic manner is needed. Sorting out individual fossil collections into communities, arraying the communities from any one time interval on a community framework, and defining community groups occurring within distinct ecologic-evolutionary units enable the geologist to systematically incorporate all the available environmental content of fossil collections, including material from wells, into the routine basin

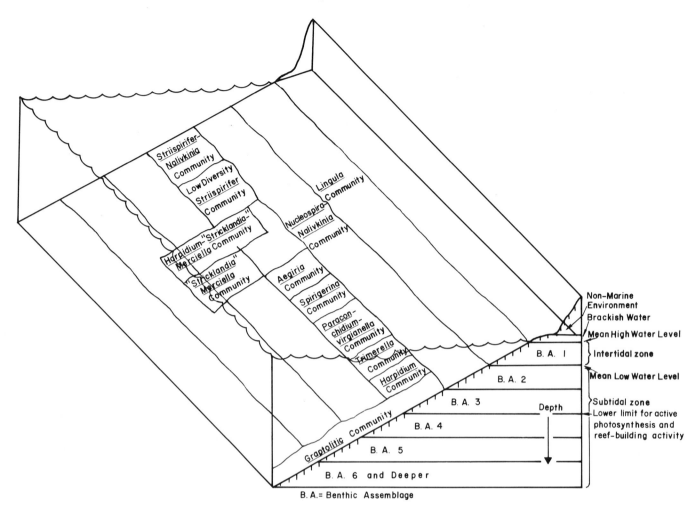

Figure 4. A block diagram, based on the data presented in Figure 10 (Community Framework for the Upper Llandoverian–lowest Wenlockian), providing a graphic view of the inferred lateral relations, as well as the shoreline proximity with depth relations of the community units discussed in the alphabetical list for this time interval.

analysis, thereby achieving a higher level of resolution. The full potential of fossils must be exploited if we are really to discharge our obligations to both science and industry.

## MATERIAL USED IN THE STUDY

Thanks to the support provided by a Senior Scholar award of the Committee on Scholarly Communication with the People's Republic of China, and the support of the Academia Sinica, Boucot spent four months during 1980 at the Nanjing Institute of Geology and Palaeontology, Academia Sinica. During this period the authors collaborated closely both in the field and in the laboratory on a variety of problems including the nature of the Chinese Ashgillian through Eifelian brachiopod-dominated communities. Most information for the Ashgillian and Silurian

came from inspection of South China material, because the Ashgillian and Silurian of North China, due to more difficult access and greater structural complexity, have not yet been analyzed in as much detail as their southern counterparts. Fresh samples were collected for paleoecologic study from the Ludlovian and Pridolian of the Qujing area, eastern Yunnan, under the leadership of Dr. Fang Run-sen, Yunnan Institute of Geological Science, Kunming. The Gedinnian through Eifelian material was taken largely from Nanjing Institute collections. Boucot spent four days collecting in the Devonian of the Liujing area, Hengxian County, southern Guangxi, under the leadership of Dr. Kuang Guo-dun, Guangxi Bureau of Geology; this trip produced a wealth of additional useful data. The far more limited North China Devonian collections available for study in Nanjing were supplemented by correspondence with a number of cooperating scientists, and also by careful consideration of the available literature. Many publications consulted are internal reports of various Chinese governmental agencies and are available only to authorized personnel.

Most of the fossil collections in the Nanjing Institute, assembled over time with attention to biostratigraphy and correlation problems, are not ecologically adequate samples. This is also true of most museum collections elsewhere in the world. Thus, the amount of community information that can be extracted from them is more limited than would be the case with fresh collections assembled primarily with community paleoecology studies in mind. However, the fresh Ashgillian-age collections made near Kuanyinchiao, the fresh Upper Silurian collections made near Qujing, and the fresh Devonian collections made near Liujing gave additional precision; conclusions based on the old museum collections were substantiated, refined, and extended by using these new collections.

To orient the taxonomist unfamiliar with many Chinese brachiopod genera, we have provided 20 plates that illustrate the bulk of the taxa involved in our study. All of the illustrated specimens are deposited in the Nanjing Institute of Geology and Palaeontology, Academia Sinica, Chi-Ming-Ssu, Nanjing. An alphabetical list of the illustrated taxa, with their positions in the plates, is also provided for the reader.

## PROCEDURES

The collections housed in Nanjing were analyzed to determine their generic composition and the relative abundance of the various taxa. Counts made of those collections that warranted attention are provided in tabular form, together with descriptions of the communities they represent. Our conclusions about what constitutes a community were determined partly by Boucot's experience with faunas of similar age elsewhere in the world. Our guiding principle throughout the analysis was to regard a community as a regularly recurring association of taxa in which the relative abundances of the various taxa are similar, within reasonable limits, but by no means absolutely the same. We considered the nature of the entombing matrix for guidance as to both environment of life and conditions of deposition where the two might not be identical. We also used available data concerning paleogeographic relations in the Chinese Ashgillian through Eifelian to aid in the environmental interpretation.

Collections made in the Devonian of the Liujing area, chiefly "handpicked" specimens, were counted by Zhang Ning, Nanjing Institute of Geology and Palaeontology, who also took part in the actual collecting. He sorted and identified the material under Rong's supervision. In addition, Zhang Ning made several bulk collections of surface material collected in bags rather than handpicked from the surface. This material was obtained from about the upper 15 cm of loose material; no effort was made to dig down to undisturbed bedrock. These bulk samples were then carefully picked in the laboratory to compare taxonomic composition and relative abundances between handpicked and bulk samples. The bulk sample contained a far more realistic represen-

tation of small specimens, and showed that the greater abundance of large specimens in the handpicked sample is largely a collecting artifact.

This work also pointed out the fact that the bulk samples collected in a reasonable length of time, and then processed later, are too small to afford an accurate count of the number of larger size taxa present in a community. Thus it is essential that both collecting procedures be employed to provide a realistic view of a community that includes both abundant (commonly smaller) taxa and rarer (commonly larger) taxa. The rarer small taxa have a better chance of being obtained from the bulk samples than from the handpicked ones. Our conclusions are in accord with earlier work carried out by Sparks (*in* Ager, 1963).

Zhang Ning also prepared a bedrock bulk sample from the Hirnantian Kuanyinchiao Formation and found it very similar in content and abundance to an earlier handcollected bedrock sample. The only difference was that the larger bulk sample, expectedly, contained more taxa than did the smaller sample.

After initial description of the community units, they were classified on community frameworks for the following major time intervals: Ashgillian, Lower Llandoverian, Middle Llandoverian, Upper Llandoverian–lowest Wenlockian, Ludlovian-Pridolian of South China, Ludlovian-Pridolian of North China, Gedinnian (northern China, Qin Ling Mountains, and northeast China), Siegenian (South China), Lower Emsian (South China, northern Vietnam), Upper Emsian (South China), and Eifelian (South China). We chose these time intervals and areas because of the nature of the available data.

## COMMUNITY EVOLUTION

Community evolution, when carefully analyzed, is an account of the changing content of community groups through time (see Boucot, 1975, 1981, 1982, 1983 for discussions). Community evolution is *not* the replacement of one community in a particular environment by another that is taxonomically unrelated; this is community replacement of one analogous community by another. The distinction is required when speculating whether level-bottom groups of Jurassic and Cretaceous rudistid bivalves were environmentally and functionally similar to level-bottom groups of other types of totally unrelated bivalves or of hermatypic corals. A general principle, recognized only in recent years (see Boucot, 1975, 1978, 1982, 1983 for accounts of the phenomenon) is that associations of genera persist through substantial intervals of geologic time (ecologic-evolutionary units of Boucot, 1983) as evolving communities inhabiting the same environment, although some of the species evolve slowly. Therefore, the evolution of the Chinese Ashgillian through Eifelian level-bottom brachiopod communities is considered here in terms of the community groups to which the communities belong. This also requires consideration of Ashgillian through Eifelian biogeography. Evolution of many of the Ashgillian communities is not discussed, because far too little is known about Ordovician communities in general. The Chinese Silurian belongs to the

late Silurian

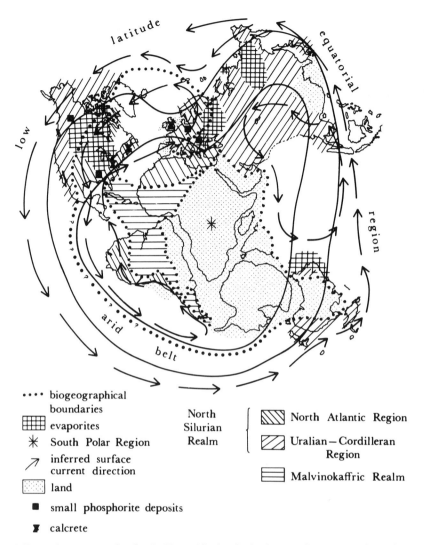

•••• biogeographical
      boundaries
▦ evaporites
✳ South Polar Region
⬈ inferred surface
      current direction
▨ land

■ small phosphorite deposits

⚑ calcrete

North
Silurian
Realm

⟨ North Atlantic Region

   Uralian–Cordilleran
   Region

   Malvinokaffric Realm

Figure 5. Pangaeic reconstruction for the Upper Silurian, indicating a surface current circulation pattern and biogeographic units. Note Arid Belt position, including location of known evaporites and calcretes. (From Boucot, A. J., 1985, Fig. 2, Philosophical Transactions of the Royal Society of London, B, v. 309, p. 326.)

Uralian-Cordilleran Region of the North Silurian Realm, rather than to the far better known North Atlantic Region. Most data in this paper come from the South China Silurian of the Uralian-Cordilleran Region (Wang Yu and others, 1984), North Silurian Realm (Fig. 5). During the Llandoverian, this region was characterized by certain genera and families that did not appear in the contemporary North Atlantic Region until about the $C_3$ part of the Upper Llandoverian, in addition to a number of strictly endemic genera only a few of which persist past the Llandoverian.

South China, during the Siegenian through Eifelian, formed a distinct Region (Fig. 6) of the Old World Realm (Wang Yu and others, 1984). We lack enough data for the South China marine Gedinnian to make any firm conclusions, although the presence of *Septoparmella* here and in northern China suggests a

lower level of endemism than later in the Devonian. During the Gedinnian, the northern Chinese region from Xinjiang (Sinkiang) through Neimongol (Inner Mongolia), and northeastern China, possibly extending into the Qin Ling Mountains and southern Gansu regions, belonged to biogeographic units that have remarkably little in common at the generic level with the South China Region.

For the northern Chinese Devonian units, we conclude that Xinjiang on the northwest may be appropriately included within the Uralian Region of the Old World Realm, whereas eastern Neimongol through northeast China (Heilongjiang, Jilin Provinces) belongs to a Mongolo-Okhotsk biogeographic subdivision of the Uralian Region. We also recognize that there may have been some boundary mixing of the South China Region with the

later early Devonian

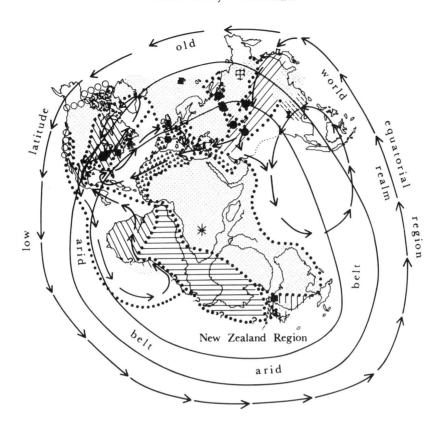

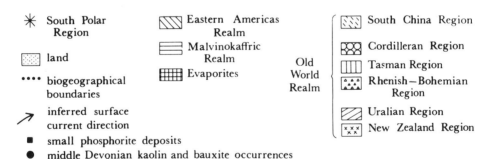

Figure 6. Later Early Devonian, Emsian biogeography on Pangaeic base (from Boucot, A. J., 1985, Fig. 1, Philosophical Transactions of the Royal Society of London, B, v. 309, p. 325.)

Mongolo-Okhotsk subdivision, as well as with the standard Uralian Region, in the Qin Ling Mountains and southern Gansu region.

Our Devonian community evolution story deals mostly with Siegenian through Eifelian brachiopods from the South China Region.

### Ordovician

As mentioned previously, there has been very little formal consideration of Ordovician communities or their community evolution. This being the case, we can do little more than provide a community framework (Fig. 7) for our Chinese data at this time.

### Silurian

In considering the evolution of the various communities, the *Eospirigerina* and *Spirigerina* Communities that belong to the *Striispirifer* Community Group (Figs. 8, 10) began outside of China in the Ashgillian of the North Silurian Realm area, and continued into the Silurian. (*Eospirigerina* does not occur above

| Benthic Assemblage | Normal Current Activity | | Rough-Water Conditions | Quiet Water, Possibly Anoxic Bottoms |
|---|---|---|---|---|
| | Medium to High Diversity | Low Diversity | | |
| 1 | | | | |
| 2 | Rhynchotrema chechiangensis Community | | | |
| 3 | Aphanomena-Hirnantia-Dalmanitina Community<br><br>Aphanomena-Hirnantia Community<br><br>"Paracraniops"-Paromalomena Community<br><br>Trimurellina-"Eoconchidium" Community | Paromalomena-Aegiromena Community | | |
| 4 | | | | |
| 5 | | | | Graptolitic |
| 6 | | | | Community |

Figure 7. Community framework for the Ashgillian of China (Hirnantian plus older beds). (Note: The following applies to Figs. 7 through 17: The significance of the vertical and horizontal ordinates and the utility of the community frameworks for basin analysis purposes are discussed on pages 2–6).

the Ashgillian outside of South China.) The *Spirigerina* Community may persist into the Lower Devonian of the Uralian and Cordilleran Regions where beds dominated by *Spirigerina* are known. The closely related *Eospirigerina and Spirigerina* behave ecologically in about the same manner.

The *Eospirigerina-Hindella* Community is closely allied to both the *Eospirigerina* and *Cryptothyrella* Communities and to the *Meristina* and *Meristella* Communities, all of which belong to the *Striispirifer* Community Group. The taxonomic lineage *Cryptothyrella-Meristina-Meristella* occurs in the high dominance, Benthic Assemblage 2, named communities, as well as in high-diversity, low-dominance communities. The genus *Hindella* is very closely related to *Cryptothyrella* (Sheehan, 1977), both morphologically and ecologically. Thus, the *Eospirigerina-Hindella* Community of the South China Lower Llandoverian (Fig. 8) may be viewed as a boundary mixture between the eospirigerinid and cryptothyrellid branches of the *Striispirifer* Community Group. All these units occur in either Benthic As-

semblage 2 or 3 (see Boucot, 1975, 1981, for Benthic Assemblage nomenclature).

The *Borealis* Community of the Middle Llandoverian (Fig. 9) belongs to the Virgianidae Community Group, and is a typical rough-water Benthic Assemblage 3 unit in Europe, although many articulated shells of *Borealis* itself occur at some localities in northeastern Guizhou, which indicates quiet-water conditions at least during the adult growth stage there. The *Borealis* Community occurs elsewhere in both the Lower and Middle Llandoverian. Its possible evolutionary relation to earlier Ashgillian Virgianidae Community Group units is unknown, but it probably is the stem community from which the *Pentamerus* Community of the Upper Llandoverian was derived. The *Pentamerus* Community in turn gave rise to the varied Pentameridae Community Group units of the later Llandoverian through Ludlovian and Pridolian times.

The *Paraconchidium–Virgianella* Community of the South China earliest Upper Llandoverian (Fig. 10) is another member

| Benthic Assemblage | Low to Medium Diversity | | Quiet Water, Possibly Anoxic Bottoms |
|---|---|---|---|
| | Moderate Current Activity | Low Current Activity, Possibly Low Oxygen | |
| 1 | | | |
| 2 | | | |
| 3 | Eospirigerina-Hindella<br>Community | Eospirigerina<br>Community | Graptolitic<br>Community |
| 4 | | | |
| 5 | | | |
| 6 | | | |

Figure 8. Community framework for the Lower Llandoverian of China.

of the Virginidae Community Group; it is presently known only from localities in northeastern Guizhou. The exact evolutionary relationship of *Paraconchidium* and other virgianids has not yet been deciphered, but clearly they are ecologically similar. Nikiforova and Sapelnikov (1971) assigned virgianid communities from the west slope of the Urals containing the similar genus *Pseudoconchidium* a Wenlockian rather than a Llandoverian age. This is surprising, as virgianids are unknown elsewhere in the world above the Llandoverian.

The *Harpidium-Stricklandia-Merciella* Community of the South China lower Upper Llandoverian is interpreted as a high-diversity, quiet-water mixture of taxa occurring near the boundaries of the Pentameridae, Stricklandiidae, and *Striispirifer* Community Groups. The quiet-water environment suggested by the sediments indicates that it had much more in common with the Stricklandiidae and *Striispirifer* Community Groups than with the Pentameridae Community Group that commonly occurs in rough water.

The *Stricklandia* Community Group *sensu strictu* has not yet been recognized in the Chinese Silurian, but the presence of the medium- to high-diversity *Stricklandia-Merciella* Community deserves comment. This is a relatively quiet-water unit, as judged by the available sedimentary evidence occurring in the Benthic Assemblage 4 position. The relative abundance of *Stricklandia* indicates an affinity with the *Stricklandia* Community Group, but the similar abundance of *Brevilamnulella* and the presence of *Pleurodium* show that this unit is also close to the *Dicoelosia-Skenidioides* and Virgianidae Community Groups.

The *Brevilamnulella* Community is similar in the South China Middle Llandoverian (Fig. 9) to occurrences in the Ashgillian and Llandoverian of the North Atlantic Region. This community was widely distributed in the North Silurian Realm and gave rise to the equally widely distributed *Clorinda* Community. Species-level relations have not yet been worked out for these units.

The *Aegiria* Community of the South China Middle Llan-

| Benthic Assemblage | Low Diversity | | | Quiet Water, Possibly Anoxic | Low Diversity, Normal Current Activity |
|---|---|---|---|---|---|
| | Rough Water | Quiet water, Possibly Low Oxygen | | | |
| 1 | | | | | |
| 2 | | | | | |
| 3 | <u>Borealis</u> Community | <u>Zygospiraella-Brachyelasma</u> Community | <u>Beitaia-Eospirifer</u> Community | Graptolitic Community | |
| 4 | | aff. <u>Anabaria</u> Community | <u>Aegiria</u> Community | | <u>Shalerid-"Dalmanella"</u> Community |
| 5 | | <u>Brevilamnulella</u> Community | | | |
| 6 | | | | | |

Figure 9. Community framework for the Middle Llandoverian of China.

doverian (Fig. 9) occurs in the Benthic Assemblage 4-5 position that is normal for a *Dicoelosia-Skenidioides* Community Group unit. However, the presence of the more offshore *Aegiria* Community in the South China, Benthic Assemblage 3, Upper Llandoverian part of the Xiushan Formation is puzzling. The *Aegiria* Community of the Xiushan Formation has been interpreted as indicating an unusual occurrence of low-oxygen conditions, or some other restrictive factor(s), that permitted this community to "invade" a nearer shore region than is normal.

Characteristic of the Uralian-Cordilleran Region Llandoverian (Figs. 8-10) are a series of communities belonging to an endemic branch (subgroups) of the *Striispirifer* Community Group. The communities are rich in costellate atrypaceans of the *Nalivkinia, Beitaia,* and aff. *Anabaria* group on the one hand, and of eospiriferids on the other. These are low- to medium-diversity communities and most occur in Benthic Assemblage 3. A few other, unrelated, genera are present, although in view of our still-limited sample this may merely represent a community

boundary mixing effect. Included here are the *Beitaia-Eospirifer* Community, *Zygospiraella-Brachyelasma* Community, aff. *Anabaria* Community, *Striispirifer-Nalivkinia* Community, *Striispirifer* Community, and *Nucleospira-Nalivkinia* Community. Not enough is known about the evolution of *Beitaia, Nalivkinia,* aff. *Anabaria,* and even some of the eospiriferids to allow interpretation of the evolution of these communities. However, they are very conspicuous within the Uralian-Cordilleran Region from at least southeastern Kazakhstan through South China, and may even occur in Tasmania.

The well-known *Atrypoidea* Community of the *Striispirifer* Community Group is present in both South China and North China (Wang Yu and others, 1980) (Figs. 11, 12). This community is characteristic of the Upper Silurian of the Uralian-Cordilleran Region where it is widespread in the Benthic Assemblage 2, quiet-water environment. The evolution of the species-level units requires additional study although Jones (1977) has made a substantial beginning.

| Benthic Assemblage | Low to Medium Diversity | | Medium to High Diversity | Low Diversity, Possibly Low Oxygen | Rough Water | Quiet Water, Possibly Anoxic Bottoms |
|---|---|---|---|---|---|---|
| | Normal Current Activity | | | | | |
| 1 | *Lingula* Community | | | | | |
| 2 | *Nucleospira*-*Nalivkinia* Community | | | | | |
| 3 | *Striispirifer*-*Nalivkinia* Community | Low Diversity *Striispirifer* Community | | *Aegiria* Community; *Spirigerina* Community | *Paraconchidium*-*Virgianella* Community; *Trimerella* Community; *Harpidium* Community | |
| 4 | | | *Harpidium*-"*Stricklandia*"-*Merciella* Community; "*Stricklandia*"-*Merciella* Community | | | |
| 5 | | | | | | *Graptolitic* Community |
| 6 | | | | | | |

Figure 10. Community framework for the Upper Llandoverian–lowest Wenlockian of China.

The widespread *Protathyris* Community, which belongs to the *Striispirifer* Community Group, of the North Silurian Realm and the Old World Realm of the Lower Devonian, is known from the Ludlovian of South China (Fig. 11). Evolutionary relations within this Community Subgroup at the species level have not been worked out. Allied to the *Protathyris* Community in South China is the *Eoschizophoria hesta-Protathyris xungmiaoensis* Community.

Within the Upper Silurian of China, *Protathyrisina* (*Protathyrisina* is similar to *Molongia*) dominates several communities; this genus is not yet recognized with certainty beyond the borders of China in the Uralian-Cordilleran Region. The communities in question are *Protathyrisina* (*plicata, minor,* and *uniplicata*), *Protathyrisina uniplicata-Striispirifer, Atrypoidea-Protathyrisina uniplicata-Striispirifer,* and *Protathyrisina-Gypidula*. The *Striispirifer-Molongia* Community is a typical medium- to high-diversity *Striispirifer* Community Group unit that occurs stratigraphically beneath a *Dicoelosia-Skenidioides* Community Group unit. Evo-

lutionary relations of all these *Striispirifer* Community Group, Benthic Assemblage 3 units, are still not well understood, and they prove to be endemic to eastern Asia.

The *Tuvaella gigantea* Community is a Benthic Assemblage 3 (Fig. 12), medium-diversity, *Striispirifer* Community Group unit, endemic to east-central Asia (Su, 1981; Rong and Zhang, 1982; Zhang and others, 1986), that features a fairly high dominance of *Tuvaella. Tuvaella* is a zygospirid (Copper, 1977a) unrelated to other Silurian genera, although seemingly derived from earlier zygospirid ancestors. *Zygospira* itself is very abundant, although not dominant, in the Benthic Assemblage 3 units of the North American Upper Ordovician (Bretsky, 1970).

The *Howellella tingi* Community of the South China Late Ludlovian-Early Pridolian (Fig. 11) is another *Striispirifer* Community Group unit that has a wide Silurian distribution in the Benthic Assemblage 2 position, and also continues into the Lower Devonian. Until the species-level evolution is understood for the many species of *Howellella,* not much can be said about

| Benthic Assemblage | Normal to Low Current Activity | | | |
|---|---|---|---|---|
| | Low Diversity | | Medium Diversity | |
| 1 | | | | |
| 2 | Atrypoidea Community | Howellella tingi Community | Protathyris Community | |
| 3 | Protathyrisina uniplicata-Striispirifer Community | Atrypoidea-Protathyrisina uniplicata-Striispirifer Community | Protathyrisina plicata, minor, uniplicata Community | Eoschizophoria hesta-Protathyris xungmiaensis Community |
| 4 | | | | |
| 5 | | | | |
| 6 | | | | |

Figure 11. Community framework for the Ludlovian-Pridolian of South China.

the evolution of their communities. However, units dominated by *Howellella* certainly occur in many places of similar environment.

Present in the Ludlovian-Pridolian of North China is a *Dicoelosia-Skenidioides* Community Group fauna about which little is known other than that it is a high-diversity community including both *Dicoelosia* and *Skenidioides*.

The Shaleriid-"*Dalmanella*" Community, another *Dicoelosia-Skenidioides* Community Group unit, was defined chiefly because it represents the oldest known occurrence of the shaleriids (the oldest record previously was in the Wenlockian of the North Atlantic Region). The sample is too small for other than recognition of community group affinity.

The Gypidulinid Community Group units are known from only one area of North China. This undoubtedly reflects a sampling artifact that will change as we learn more about the Chinese Silurian, although the rarity of shelly Wenlockian and younger Silurian in South China may limit the occurrences almost entirely to North China.

## Devonian

We first discuss those community group units that are sparsely represented in terms of numbers of communities in the Chinese Gedinnian through Eifelian.

The Gypidulidae Community Group is represented only by the *Zdimir* Community in South China (Fig. 16). However, this is a globally widespread community unit with which the South China material shares all major characteristics, i.e., high-dominance, low-diversity, mass aggregations rich in pedicle valves, and a rough-water Benthic Assemblage 3 situation. Not enough is known about the species-level evolution of *Zdimir* to say anything of consequence, but it is clear that, throughout the later Early Devonian and the Middle Devonian, no changes of environmental-ecological significance took place within the genus.

Bivalve, receptaculitid, vertebrate, and *Nowakia* Community group types occur in ecologically appropriate positions within

| Benthic Assemblage | Normal Current Activity | | Low Diversity | |
|---|---|---|---|---|
| | Medium Diversity | High Diversity | Quiet Water | Rough Water |
| 1 | | | | |
| 2 | | | Atrypoidea Community | |
| 3 | Tuvaella gigantea Community | Striispirifer- Molongia Community | Protathyrisina- Gypidula Community | Gypidulid- Uncinulid Community |
| 4 | | Dicoelosia- Skenidioides Community Group | | |
| 5 | | | | |
| 6 | | | | |

Figure 12. Community framework for the Ludlovian and Pridolian of northern China.

the South China Devonian; they substantiate conclusions made elsewhere about their environmental fixity through time. The bivalve-dominated community occurs close to the shoreline region, as does the vertebrate-dominated community (or communities, as the vertebrates have not really been given close ecologic scrutiny; see Blieck, 1982, for what can be accomplished with vertebrate communities). The receptaculitid community type occurs in Benthic Assemblage 3, and its environment in South China appears similar to that reported elsewhere for that community type in the earlier Paleozoic. The *Nowakia* communities also occur in thin-bedded to laminated strata similar to those in which they occur elsewhere in the world. In China, as elsewhere, they are interpreted as "basinal" facies of a "deep-water" type associated with a chiefly pelagic fauna, and little or very restricted benthos except at their extreme margins. Most of the benthos seems to belong to the mid- to outer shelf *Dicoelosia-Skenidioides* Community Group.

The *Dicoelosia-Skenidioides* Community Group units are not widespread for the time interval considered here (Fig. 13), except in the Nandan Facies of southern China. The Nandan Facies is a characteristic basinal facies of the mid-Paleozoic featuring abundant *Nowakia* and goniatites, rare shelly benthos, and a few trilobites. The brachiopod communities consist chiefly of very scattered small shells, as is characteristic of the *Dicoelosia-Skenidioides* Community Group. The Nandan fauna has much in common with Langenstrassen's (1972) Fredeberger Schichten. Most of our knowledge of these communities in South China comes from Xu (1977, 1979). Xu, who has reported (oral communication, 1980) that these brachiopods occur as very low population density aggregations, has done an excellent job of describing the available material. As more material becomes available, we expect our understanding of the Nandan Facies brachiopod communities to be considerably improved. At present the Nandan Facies types include *Buchiola-Reticulariopsis* Community (Late Emsian-Eifelian), *Cryptatrypa-Strophochonetes* Community (Emsian), *Plectodonta-Reticulariopsis* Community

| Benthic Assemblage | Quiet Water | | |
| --- | --- | --- | --- |
| | Medium to High Diversity | Low Diversity | |
| 1 | | | |
| 2 | *Protathyris-Lanceomyonia* Community | | |
| 3 | | | |
| 4 | *Septoparmella* containing *Dicoelosia-Skenidioides* Community Group | *Septoparmella* Community | *Septoparmella-Aldanispirifer* Community |
| 5 | | | |
| 6 | | | |

Figure 13. Community framework for the Gedinnian of northern China (Qin Ling Mountains and northeastern China).

(Eifelian), and *Reticulariopsis-"Chonetes"* n. sp. Community (Eifelian). Three of these communities have relatively abundant *Reticulariopsis. Reticulariopsis* is not present in the Australian earlier Devonian beds containing the somewhat similar *Maoristrophia* Community faunas, but the overall similarity, especially the presence of notanopliids and a few other taxa, is important. *Dicoelosia* and *Skenidioides,* or descendant taxa, have not yet been found in the Nandan Facies brachiopod communities, but presumably they will be found when larger collections become available. The amalgamation of the pelagic *Nowakia* Community fossils with the benthic shells is also noteworthy. It is conceptually important that pelagic community units be kept distinct from benthic units. In terms of community evolution, not enough information has accumulated about the Nandan Facies *Dicoelosia-Skenidioides* Community Group units to allow any significant conclusions; data from future collections may alter this situation markedly.

Also important to the *Dicoelosia-Skenidioides* Community

Group story is the *Septoparmella*-containing *Dicoelosia-Skenidioides* Community Group material from the Erdaogou Formation in Jilin Province, northeastern China, and the *Septoparmella*-spiriferid communities of both North and South China (Fig. 13). *Septoparmella* is another notanopliid, but in this highly endemic group we find that this Uralian, as well as South China Region Gedinnian representative, is markedly different ecologically from the three younger genera endemic in the South China Region. *Septoparmella* occurs in a wide swath in east-central Asia from southeastern Kazakhstan to Jilin Province, i.e., in the Mongolo-Okhotsk subdivision (either a Subprovince or a Province) of the Uralian Region, and in Guangdong on the south. Some Chinese *Septoparmella* also occurs in nearer-shore communities than do other notanopliids, reaching into the Benthic Assemblage 3 position. Yet *Septoparmella* in the Gedinnian Erdaogou Formation, together with other taxa compatible with a Benthic Assemblage 4-5 *Dicoelosia-Skenidioides* Community Group fauna, indicates that the notanopliid depth range in China

| Benthic Assemblage | Quiet Water | | Normal Current Activity | | |
|---|---|---|---|---|---|
| | | Lower Oxygen | Low Diversity | | |
| 1 | | | Bivalve Community | Vertebrate Community | |
| 2 | | Protathyris Community | | Orientospirifer-Sinochonetes Community | |
| 3 | | | | | |
| 4 | | | | | |
| 5 | | | | | |
| 6 | | | | | |

Figure 14. Community framework for the Siegenian of South China.

is similar to that suggested elsewhere. *Septoparmella* in South China is consistent with a *Dicoelosia-Skenidioides* Community Group assignment. In North China, the Neimongol *Septoparmella-Aldanispirifer* Community may also belong to the *Dicoelosia-Skenidioides* Community Group.

Most of the Siegenian through Eifelian South China Region communities belong to the *Striispirifer* Community Group. The Mongolo-Okhotsk subdivision *Sinostrophia-Discomyorthis* Community, based on information provided by Hamada (1971) and the *Fallaxispirifer-Discomyorthis* and *Coelospirella* Communities, based on information provided by Su Yang-zheng (written communication, 1980), are typical high-diversity, very endemic *Striispirifer* Community Group units from the Benthic Assemblage 3 region. They are parallel developments, reflecting the high provincialism of Emsian time, of the Eastern Americas Realm *Amphigenia* Community, the Tasman Region *Quadrithyris* Community, and varied communities from the Rhenish Complex of Communities in the Rhenish-Bohemian Region. When

more is learned about the distribution and occurrence of these communities in northern China, it may be possible to subdivide them appropriately. In any event, they are normal environmemt, muddy sand substrate, Benthic Assemblage 3 units, despite the curious biogeographic mixing represented by the Old World Realm cosmopolitan taxa, Eastern Americas Realm, and Tasman Region vagrants, and purely endemic taxa characteristic of the Mongolo-Okhotsk subdivision. The *Coelospirella* Community is a low-diversity, high-dominance Benthic Assemblage 3 unit similar to the Silurian *Coelospira* Community.

The Benthic Assemblage 2, *Striispirifer* Community Group units in South China are normally of much lower diversity than those in Benthic Assemblage 3. The Gedinnian *Protathyris-Lanceomyonia* Community from southeastern Gansu (Fig. 13) is a typical low-diversity, somewhat quiet-water *Striispirifer* Community Group unit dominated by a rostrospiroid and a rhynchonellid. *Protathyris* and various rhynchonellid genera of the Lower Devonian are common in Benthic Assemblage 2, in high-

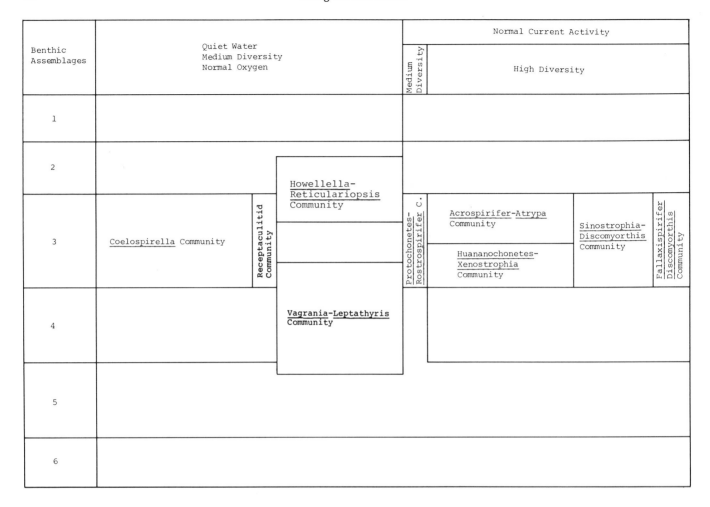

Figure 15. Community framework for the Lower Emsian of South China.

dominance, low-diversity units that probably represent different sets of restrictive conditions. The *Protathyris* Community present in the South China Siegenian is a good example (Fig. 14). The associated *Orientospirifer-Sinochonetes* Community is another of these low-diversity, high-dominance units. In low-diversity, high-dominance Benthic Assemblage 2, *Striispirifer* Community Group units involving dominant chonetid taxa, relatively turbid conditions may be the restrictive factor. The *Howellella-Reticulariopsis* Community of the Lower Emsian in South China is another of the low-diversity, high-dominance units so characteristic of the *Striispirifer* Community Group in the Benthic Assemblage 2 or 3 position (Fig. 15). The factors responsible for the various permutations and combinations represented by the taxic dominants are still only poorly understood. This first step of sorting out the community units will ultimately enable us to understand why the units are present or absent from place to place. The last Benthic Assemblage 2 unit dealt with is the Eifelian *Rhipidothyris* Community (Fig. 17). This low-diversity community, about which we know little in terms of associated communities, is similar to the ancestral *Globithyris* and *Rhenorensselaeria* Communities. This suggests a quiet-water, very restrictive, Benthic Assemblage 2 environment.

The remaining *Striispirifer* Community Group units (from the Benthic Assemblage 3 or 3-4 position) may be grouped into three broad classes. The first involves an abundance or even dominance of protochonetids of one sort or another (*Protochonetes-Rostrospirifer* Community, *Huananochonetes-Xenostrophia* Community) that may represent more turbid water. The second involves an abundance or even dominance of large spiriferids (*Acrospirifer-Atrypa* Community, *"Subcuspidella"-Athyrisina* Community, *"Euryspirifer" qijianensis-Kwangsia* Community, *Rostrospirifer-Athyrisina* Community, *"Acrospirifer" fongi–Eospiriferina lachrymosa* Community, *Atrypa-Xystostrophia* Community) that may indicate less turbid, relatively normal conditions. The third is the *Vagrania-Leptathyris* Community, which involves a different group of taxa and may represent quite different, more offshore, conditions from the first two.

| Benthic Assemblages | Quiet Water Medium Diversity | | | Normal Current Activity High Diversity | Rough Water |
|---|---|---|---|---|---|
| | Low Oxygen | 1 | Normal Oxygen | | |
| 1 | | | | | |
| 2 | | | | | |
| 3 | | | "Subcuspidella"-Athyrisina Community  Rostrospirifer-Athyrisina Community | "Euryspirifer" qijianensis-Kwangsia Community | Zdimir Community |
| 4 | Buchiola-Reticulariopsis Community | Cryptatrypa-Strophochonetes Community | Nowakia Community | | |
| 5 | | | | | |
| 6 | 1 Possible anoxic bottom | | | | |

Figure 16. Community framework for the Upper Emsian of South China.

## CONCLUSIONS

The Chinese Silurian-Devonian brachiopod communities provide additional support for concepts of community evolution developed largely from non-Chinese data (Boucot, 1978). Many of the Uralian-Cordilleran Region Silurian communities (see Wang Yu and others, 1984, for biogeographic data) have Chinese counterparts that are similarly related to their environments, stratigraphic position, and to other communities (compare community framework diagrams in this paper with those for the Silurian in Boucot [1975, particularly Figs. 3-5]), and for the Devonian (Boucot, 1982). In addition, some of the endemic Llandoverian faunas of the Uralian-Cordilleran Region are analyzed, for the first time, in terms of communities. Many of these communities are endemic, and unknown in the North Atlantic Region, as would have been predicted from knowledge of Silurian historical biogeography. Evolution under conditions that promoted biogeographic isolation, long-term reproductive isolation giving rise to community subgroups, has not materially affected the community group characteristics.

We conclude that larval behavior of these sessile benthic organisms has been very conservative through geologic-evolutionary time, as contrasted with genus- and species-level morphologic changes. We emphasize that family-level evolutionary changes do not appear to have been involved in any of these community groups. No gradual community changes took place that might have paralleled family-level morphologic changes. During the Ashgillian-Eifelian interval, groups of families came and went, as did community groups, but we see no evidence for gradual change in the larval behavior of any of the taxa within a community group. This does not mean that all taxa within a community group possess precisely the same environmental tolerances. Some are clearly far more eurytopic than others, as shown by their presence in more than one community group, or in more than one community subgroup branch of a community group.

The next logical step in our work, beyond the need for collecting more data of the kind presented here, is more detailed sampling. Most of these Chinese community units were examined only in a reconnaissance way. For the most part, they are based

| Benthic Assemblage | Quiet Water Low Diversity | | Normal Current Activity High Diversity | | Rough Water Low Diversity |
|---|---|---|---|---|---|
| | Low Oxygen | Possible Anoxic Bottom | | | |
| 1 | | | | | |
| 2 | | | Rhipidothyris Community | | |
| 3 | | | | "Acrospirifer" fongi-Eospiriferina lachrymosa Community | Zdimir Community |
| 4 | Plectodonta-Reticulari-opsis Community | Nowakia Community | Leiorhynchus Community | Reticulari-opsis-"Chonetes" n. sp. Community | |
| 5 | | | | | |
| 6 | | | | | |

Figure 17. Community framework for the Eifelian of South China.

on handpicked collections whose fossils are derived from more than a single, taxonomically homogeneous layer. Of course, we recognize that many layers are not laterally homogeneous, particularly if one moves far enough from one collecting site to another. We must sample closely spaced, taxonomically homogeneous intervals, at the same time paying careful attention to their lithologic character. We must also search more diligently for both sedimentary structures and trace fossils that will provide additional environmental clues. After this detailed work is carried out, we will much better understand why similar or different associations occur in what at first appear to be similar rocks, and also what at first glance appear to be dissimilar rocks. Obviously we cannot do this type of detailed sampling and study over wide regions. Nevertheless, it is critical that effective paleoecologic reconnaissance studies be made to recognize the problems before undertaking the far more laborious, time-consuming, detailed paleoecologic work.

We believe that research in the community paleoecology of fossil benthos has great potential for providing a better understanding of past environments, as an aid in basin analysis, and for understanding the behavior of the organisms through evolutionary time. Results thus far are consistent with a concept of long-term stabilizing selection, interrupted by brief intervals of marked change and reorganization that coincide with marked taxonomic changes at the family and higher levels.

## NAMING COMMUNITIES

Community names have been devised by biologists to indicate the one or two more common genera and their species. In effect, however, this approach to nomenclature results in naming geologically demonstrable community groups. For those concerned with the evolutionary consequences of community evolution, such a system of naming communities is unsatisfactory. If one is concerned with discriminating species-level changes that take place during the evolution of a community, clearly most of the specific changes take place in the less common, less abundant genera (Boucot, 1978). These are species of the less abundant genera that are ignored by the biologist naming modern communities. The paleontologist, too, commonly has used the abundant,

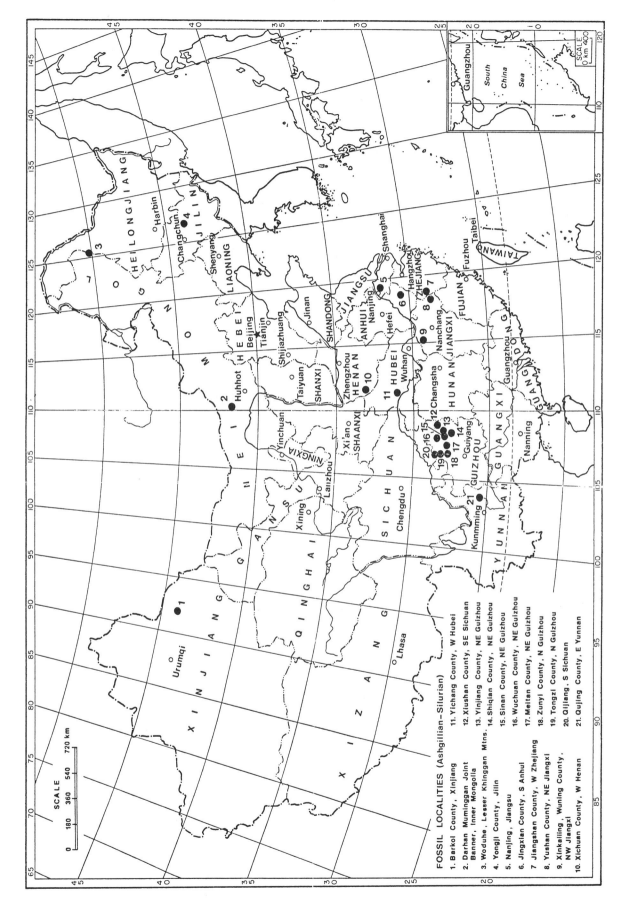

FOSSIL LOCALITIES (Ashgillian–Silurian)

1. Barkol County, Xinjiang
2. Darhan Muminggan Joint Banner, Inner Mongolia
3. Woduhe, Lesser Khinggan Mtns.
4. Yongji County, Jilin
5. Nanjing, Jiangsu
6. Jingxian County, S Anhui
7. Jiangshan County, W Zhejiang
8. Yushan County, NE Jiangxi
9. Xinkailing, Wuning County, NW Jiangxi
10. Xichuan County, W Henan
11. Yichang County, W Hubei
12. Xiushan County, SE Sichuan
13. Yinjiang County, NE Guizhou
14. Shiqian County, NE Guizhou
15. Sinan County, NE Guizhou
16. Wuchuan County, NE Guizhou
17. Meitan County, NE Guizhou
18. Zunyi County, N Guizhou
19. Tongzi County, N Guizhou
20. Qijiang, S Sichuan
21. Qujing County, E Yunnan

Figure 18. Index map showing Chinese localities referred to in text.

TABLE 1. <u>ACROSPIRIFER-ATRYPA</u> COMMUNITY

| Name | Articulated Shell | Pedicle Valve | Brachial Valve | Either Valve | Total No. Specimens | % |
|------|------------------|---------------|----------------|--------------|---------------------|---|
| **Locality ADH 11** | | | | | | |
| Atrypa | 29 | | | | 29 | 30 |
| Acrospirifer papaeoensis | 15 | | | | 15 | 15 |
| Schizophoria | 4 | | | | 4 | 4 |
| Xenostrophia | 3 | | | | 3 | 3 |
| Uncinulus fasciger | 11 | | | | 11 | 12 |
| "Megastrophia" | 1 | | | | 1 | 1 |
| Athyris | 3 | | | | 3 | 3 |
| Glyptospirifer sp. nov | 4 | | | | 4 | 4 |
| Dicoelostrophia | 12 | | | | 12 | 13 |
| "Chonetes" n. gen., small, coarse | 9 | | | | 9 | 9 |
| Parachonetes | 5 | | | | 5 | 5 |
|    Total | | | | | 96 | |
| **Locality ADH 12** | | | | | | |
| Atrypa | 79 | | | | 79 | 18 |
| Acrospirifer | 84 | | | | 84 | 19 |
| Athyris | 12 | | | | 12 | 3 |
| Leptaenopyxis | 1 | | | | 1 | - |
| Schizophoria | 29 | | | | 29 | 7 |
| Rostrospirifer | 59 | | | | 59 | 13 |
| Dicoelostrophia | 12 | | | | 12 | 3 |
| Elymospirifer | 4 | | | | 4 | 1 |
| "Desquamatia" | 3 | | | | 3 | 1 |
| "Chonetes" n. gen., small, coarse | 5 | | | | 5 | 1 |
| Parachonetes | 33 | | | | 33 | 8 |
| Uncinulus | 105 | | | | 105 | 24 |
| "Areostrophia" | 1 | | | | 1 | - |
| Xenostrophia | 9 | | | | 9 | 2 |
| "Megastrophia" | 1 | | | | 1 | - |
| Glyptospirifer | 5 | | | | 5 | 1 |
|    Total | | | | | 442 | |
| **Locality ADH 12-13** | | | | | | |
| Atrypa | 3 | | | | 3 | 43 |
| Athyris | 3 | | | | 3 | 43 |
| Glyptospirifer | 1 | | | | 1 | 14 |
|    Total | | | | | 7 | |
| **Locality ADH 13** | | | | | | |
| Atrypa | 32 | | | | 32 | 14 |
| Acrospirifer medius | 46 | 3 | | | 49 | 21 |
| Athyris grandis | 4 | | | | 4 | 2 |
| Dicoelostrophia sp. | 7 | | 2 | | 9 | 4 |
| Schizophoria sp. | 73 | 5 | | | 78 | 34 |
| "Cymostrophia" | 1 | | | | 1 | - |
| Xenostrophia | 13 | | | | 13 | 6 |
| Parachonetes | 19 | | | | 19 | 8 |
| "Chonetes" n. gen., small, coarse | 2 | | | | 2 | 1 |
| Uncinulus fasciger | 16 | | | | 16 | 7 |
| Elymospirifer | 6 | | | | 6 | 3 |
| Glyptospirifer chui | 1 | | | | 1 | - |
|    Total | | | | | 230 | |
| **Locality ADH 14** | | | | | | |
| Atrypa | 2 | | | | 2 | 1 |
| Acrospirifer medius | 34 | 5 | | | 39 | 23 |
| Schizophoria | 3 | | | | 3 | 2 |
| Leptaenopyxis | 7 | | | | 7 | 4 |
| Dicoelostrophia | 3 | 3 | | | 6 | 4 |
| "Megastrophia" | 1 | | | | 1 | 1 |
| Xenostrophia | 1 | | | | 1 | 1 |
| "Chonetes" n. gen., small, coarse | 1 | | | | 1 | 1 |
| Protochonetes | | 25 | 8 | | 33 | 20 |
| Uncinulus fasciger | 55 | | | | 55 | 33 |
| Glyptospirifer chui | 10 | | | | 10 | 6 |
| Elymospirifer | 1 | | | | 1 | 1 |
| Athyris grandis | 8 | | | | 8 | 5 |
|    Total | | | | | 167 | |

TABLE 1. <u>ACROSPIRIFER-ATRYPA</u> COMMUNITY (continued)

| Name | Articulated Shell | Pedicle Valve | Brachial Valve | Either Valve | Total No. Specimens | % |
|---|---|---|---|---|---|---|
| **Locality ADH 15** | | | | | | |
| Acrospirifer | 36 | 1 | | | 37 | 46 |
| Elymospirifer | 1 | 1 | | | 2 | 3 |
| Glyptospirifer | 4 | | | | 4 | 5 |
| Leptaenopyxis | 10 | | | | 10 | 12 |
| Schizophoria | 2 | | | | 2 | 3 |
| Protochonetes | | 5 | 1 | | 6 | 7 |
| Xenostrophia | 3 | | | | 3 | 4 |
| Dicoelostrophia | 1 | 2 | | | 3 | 4 |
| Uncinulus | 1 | | | | 1 | 1 |
| "Megastrophia" | 1 | | | | 1 | 1 |
| Athyris grandis | 1 | | | | 1 | 1 |
| Levenea | 11 | | | | 11 | 14 |
| Total | | | | | 81 | |
| **Wang-Rong 1978-1979 collection** | | | | | | |
| Acrospirifer | 97 | | | | 97 | 27 |
| Rostrospirifer | 24 | | | | 24 | 7 |
| Atrypa | 54 | | | | 54 | 15 |
| Athyris | 19 | | | | 19 | 5 |
| Schizophoria | 63 | | | | 63 | 17 |
| Uncinulus | 40 | | | | 40 | 11 |
| Dicoelostrophia | 21 | | | | 21 | 6 |
| Elymospirifer | 11 | | | | 11 | 3 |
| Parachonetes | 9 | | | | 9 | 3 |
| Glyptospirifer | 6 | | | | 6 | 2 |
| Xenostrophia | 4 | | | | 4 | 1 |
| "Chonetes" n. gen., small, coarse | 3 | | | | 3 | 1 |
| "Megastrophia" | 8 | | | | 8 | 2 |
| Parathyrisina | 2 | | | | 2 | 1 |
| "Desquamatia" | 1 | | | | 1 | - |
| Total | | | | | 362 | |

dominant genera and their species when naming communities, following the lead provided by the biologists. An evolutionary approach to naming communities would, on the contrary, name them after the rapidly evolving taxa. But this is done only after enough attention has been given to any particular community group through time to study the less common, more rapidly evolving genera and their species. At this more advanced level, community names indicate less common genera and their more rapidly evolving species, whereas the more common genera and their less rapidly evolving species are reserved for community group names.

## ALPHABETICAL LIST OF COMMUNITIES

*Acrospirifer–Atrypa Community* (Fig. 15; Table 1)

*Name:* This community is named here.

*Age:* The Shizhou Member of the Yukiang Formation is of Early Emsian age (Ruan and others, 1979; Wang Cheng-yuan and others, 1979).

*Composition:* Table 1 provides data from six samples of this community, plus an inadequate seventh sample. Although the dominant taxa are largely the same ones from collection to collection, some changes in abundance rank take place. The strati-graphically highest sample, number ADH 15, is somewhat intermediate to the *Huananochonetes-Xenostrophia* Community, compositionally as well as in stratigraphic position.

*Typical Locality:* Liujing, Hengxian County, southern Guangxi.

*Geographic Distribution:* South China, including Guangxi, north Sichuan (Wang and others, 1978; Chen, 1979), eastern Yunnan (Yin, 1938), and possibly northern Vietnam, if the material published by Mansuy (1908) and Patte (1926) is proven to belong here.

*Environment:* The commonly articulated nature of the brachiopods, combined with the muddy matrix, suggests a relatively quiet-water environment. The high diversity is consistent with a Benthic Assemblage 3 assignment, as is the position above Benthic Assemblage 2 communities and below another Benthic Assemblage 3 unit. This community may represent a somewhat nearer-shore Benthic Assemblage 3 unit than does the overlying *Huananochonetes-Xenostrophia* Community. The differences between the two are probably related to some factor that correlates with shoreline proximity and depth. Total diversity of samples ADH 11-15 correlates positively with sample size, as expected.

*Community Group: Striispirifer.*

## *"Acrospirifer" fongi–Eospiriferina lachrymosa* Community (Fig. 17)

*Name:* This community is named here.

*Age:* Early Eifelian (Wang Yu and others, 1979; Wang Cheng-yuan and others, 1979) based on dacryoconarids and conodonts.

*Composition: Eospiriferina, "Acrospirifer," Glyptospirifer, Kwangsia, "Chonetes"* n. gen., productid, with the first two being most abundant. The collection available for study contains more than 100 specimens, which makes it likely that some of the rarer items are unknown.

*Typical Locality.* Yintang, Dale, Xiangzhou County, central Guangxi, in the lower part of the Yintang Formation.

*Geographic Distribution:* Guizhou, Guangxi, and Sichuan (see Wang and Zhu, 1979; Hou and Xian, 1975; Yue and Bai, 1978; Wang and others, 1978).

*Environment:* Some of the shells occur as articulated specimens in a calcareous mudstone matrix, whereas others occur as disarticulated valves in a calcareous matrix. This is probably a medium-diversity, quiet-water, Benthic Assemblage 3 community, but the restrictive factor lowering the diversity is uncertain.

*Community Group: Striispirifer.*

## *Aegiria* Community (Fig. 9)

*Name:* This community is named here.

*Composition:* Dominant small specimens of *Aegiria.* The Sichuan occurrence includes less common specimens of *Nucleospira, Septatrypa,* and *Coronocephalus,* whereas the Henan occurrence yields only *Aegiria.*

*Age:* Late Llandoverian for the Sichuan occurrence (Ge and others, 1979). Middle Llandoverian for the Henan occurrence (associated with the graptolites *Rastrites* and *Demirastrites triangularis;* Yan Guo-shun, oral communication, 1980).

*Typical Locality:* Rongxi, Xiushan County, southeastern Sichuan.

*Geographic Distribution:* Four localities in northeastern Guizhou, and southeastern Sichuan. The Henan locality is at Zhangwan, Xichuan County, Henan (Yan Guo-shun, oral communication, 1980).

*Environment:* Occurs as thoroughly disarticulated shells strewn over bedding surfaces of thin-bedded, calcareous, gray siltstone in the upper part of the Xiushan Formation. Environmentally this community presents problems. Dominance of small *Aegiria* is consistent with low-oxygen Benthic Assemblage 5 or even 6 environment, as is suggested by some non-Chinese occurrences. However, the adjacent communities are overwhelmingly Benthic Assemblage 3, suggesting that this occurrence may represent low-oxygen conditions in shallower water. Intermittent current activity was sufficient to disarticulate the shells.

The Henan specimens collected by Dr. Yan consist of disarticulated, scattered shells in a thin-bedded, calcareous mudstone matrix consistent with fairly quiet-water conditions. The associated graptolites are consistent with Benthic Assemblage 4-5 conditions. The occurrence of a Shaleriid-*"Dalmanella"* Community, *Dicoelosia-Skenidioides* Community Group, faunule 200 m below the *Aegiria* Community is consistent with a Benthic Assemblage 4 assignment.

*Community Group: Dicoelosia-Skenidioides.*

## aff. *Anabaria* Community (Fig. 9)

*Name:* This community is named here.

*Composition:* Articulated specimens of aff. *Anabaria* (this material is being described as a new genus by Dr. Yan; oral communication, 1980).

*Age:* Middle Llandoverian (see Shaleriid-*"Dalmanella"* Community); occurs at the same horizon as the Shaleriid-*"Dalmanella"* Community.

*Typical Locality:* Shiyanhe, Xichuan County, Henan, in the easternmost Qin Ling Mountains.

*Geographic Distribution:* Same as above.

*Environment:* A quiet-water environment is inferred from the presence of articulated specimens possibly representing Benthic Assemblage 4-5 as it occurs at the same horizon as the Benthic Assemblage 4-5 Shaleriid-*"Dalmanella"* Community. This community occurs about 50-60 km away from the co-eval Shaleriid-*"Dalmanella"* Community, suggesting that a Benthic Assemblage 3 assignment is possible when dealing with a new genus from such a poorly known region.

*Community Group: Striispirifer.*

## *Aphanomena–Hirnantia* Community (Fig. 7; Table 2)

*Name:* Rong (1979) used the term *Kinnella-Plectothyrella* Assemblage for what is here termed the *Aphanomena-Hirnantia* Community, although ultimately this may prove more useful as a community subgroup name.

*Composition:* Rong (1979) included the following brachiopod genera in the community: *Hirnantia, Cliftonia, Kinnella, Dalmanella, Aphanomena, Hindella, Plectothyrella,* and *Paromalomena.* Table 2 shows the relative abundances in a typical sample of this community.

TABLE 2. APHANOMENA-HIRNANTIA COMMUNITY

| Name | Articulated Shell | Pedicle Valve | Brachial Valve | Either Valve | Total No. Specimens | % |
|---|---|---|---|---|---|---|
| **Locality 200 AAE** | | | | | | |
| Aphanomena | | 11 | 9 | 32 | 52 | 37 |
| Hindella | 1 | 10 | 5 | 1 | 17 | 12 |
| Hirnantia | | 21 | 19 | 5 | 45 | 32 |
| Cliftonia | | 8 | 2 | 1 | 11 | 8 |
| Paromalomena | | 1 | 1 | 1 | 3 | 2 |
| Dalmanella? | | 5 | 2 | 2 | 9 | 6 |
| Plectothyrella | | 3 | | | 3 | 2 |
| Total | | | | | 140 | 99 |
| **Locality 200 AAE*** | | | | | | |
| Hirnantia magna | 2 | 37 | 17 | 24 | 80 | 8 |
| H. sagittifera fecunda | 3 | 77 | 35 | 8 | 123 | 13 |
| Kinnella kielanae | 1 | 1 | 11 | | 14 | 1 |
| Dalmanella testudinaria | 3 | 137 | 69 | | 209 | 21 |
| Aphanomena ultrix | | 135 | 50 | 76 | 261 | 27 |
| Hindella crassa incipiens | 20 | 54 | 17 | 7 | 98 | 10 |
| Plectothyrella crassicosta | 6 | 14 | 14 | 3 | 37 | 4 |
| Cliftonia | 3 | 65 | 8 | 9 | 85 | 9 |
| Leptaenopoma | 1 | 16 | 7 | 18 | 42 | 4 |
| Catazyga | | | 1 | | 1 | |
| Triplesia | | | 12 | | 12 | 1 |
| Dorytreta | | 2 | 12 | | 14 | 1 |
| Atrypacean | | | 4 | | 4 | |
| Philhedra | | | 3 | | 3 | |
| Lingula | | | | 2 | 2 | |
| Total | | | | | 985 | 99 |

*Sample collected in 1980; prepared and identified by Zhang Ning in 1981 at Corvallis, Oregon.

*Age:* Rong (1979) recorded this unit only from Hirnantian age, i.e., Late Ashgillian, strata of *Diceratograptus mirus, Paraorthograptus uniformis,* and *Diplograptus bohemicus* Zone ages.

*Typical Locality:* Rong proposed here that the Kuanyinchiao Bed at Guanyinqiao, Jijiang, Sichuan Province (Locality 200 AAE) be designated the typical locality.

*Geographic Distribution:* Rong (1979) recognized this community in about 30 occurrences in the following provinces: Yunnan, Guizhou, Sichuan, and Hubei, in South China.

*Environment:* The Kuanyinchiao Bed is about 50 cm thick, and tends to be very massive. The individual brachiopods occur as thoroughly disarticulated, well-preserved shells scattered within the calcareous mudstone matrix. The occurrence is not dense enough to be characterized as a shell bed. Specimens on the same slab commonly are oriented with the convex side up, suggesting some current activity. The rock is very fine grained, black, calcareous mudstone (Rong, 1979, referred to the rock as "marl"). The nature of the black pigment has not been determined, but it could be free carbon, pyrite, or both.

The above data permit us to conclude that this medium-diversity fauna represents moderate depth, possibly Benthic Assemblage 3, because it occurs landward of both the Pelagic Community (dominated in this region by graptolites) and the more offshore *"Paracraniops"-Paromalomena* Community, and is itself interpreted to be seaward of the very low diversity *Paromalomena-Aegiromena* Community. The highly disarticulated, well-sorted brachiopods, as well as their commonly convex upward orientation and scattered distribution, are consistent with a moderately turbulent, well-oxygenated environment. The absence of corals, stromatoporoids, trilobites, and bryozoans may be a function of turbid water, possibly cooler water, or a combination of both. The very calcareous nature of the rock suggests that this was not truly a cold-water environment. In both the underlying Ordovician and overlying Silurian, the local presence of abundant carbonate rocks and extra-Malvinokaffric faunas is consistent with the hypothesis that these Hirnantian faunas of South China represent a northward pulse of cooler water correlated with Southern Hemisphere glaciation during the Ashgillian.

The moderately large size of many specimens is consistent with a Benthic Assemblage 3 or 4—rather than 5 or 6—position. A modest percentage of the meristellid genus *Hindella* is also consistent with this conclusion because dominant *Hindella,* as well as the closely related *Cryptothyrella,* make up both the Benthic Assemblage 2 *Hindella* and *Cryptothyrella* Communities, and the *Meristina* and *Meristella* Communities of the later Silurian and Lower Devonian in Benthic Assemblage 2.

*Community Group:* We cannot yet assign most Ordovician communities to a community group.

TABLE 3. APHANOMENA-HIRNANTIA-DALMANITINA COMMUNITY

| Name | Articulated Shell | Pedicle Valve | Brachial Valve | Either Valve | Total No. Specimens | % |
|---|---|---|---|---|---|---|
| **Locality AAE 382** | | | | | | |
| Hindella | 3 | | | | 3 | 1 |
| Dorytreta | | 1 | 1 | | 2 | 1 |
| Aphanomena | | 12 | 11 | 87 | 110 | 46 |
| Fardenia | | 1 | | 4 | 5 | 2 |
| Hirnantia | | 45 | 32 | 11 | 88 | 37 |
| Leptaenopoma | | 8 | 8 | 8 | 24 | 10 |
| Leptaena | | | 1 | 4 | 5 | 2 |
| Total | | | | | 237 | |

| | Pygidium | Cephalon | Either | Total No. Specimens |
|---|---|---|---|---|
| Dalmanitina | 26 | 32 | 4 | 62 |
| Trochonema | | | − | − |
| Cornulites | | | 10 | 10 |
| Cypricardinia | | | 2 | 2 |
| Goniophora | | | 1 | 1 |
| Sphenotus | | | 1 | 1 |
| Total | | | | 313 |

*Aphanomena–Hirnantia–Dalmanitina* Community (Fig. 7; Table 3)

*Name:* This community is named here.

*Composition:* The taxa in the community are listed in Table 3, which also indicates their relative abundances at the typical locality. The specimens of *Hirnantia* average more than twice the size of those present in the allied *Aphanomena-Hirnantia* Community. Noteworthy is an abundance of the trilobite *Dalmanitina*, uncommon bivalves, a gastropod, *Cornulites,* and pelmatozoan debris, normally present in the *Aphanomena-Hirnantia* Community.

*Age:* Rong (1979) assigned the typical locality (AAE 382) to the late Ashgillian.

*Typical Locality:* Locality AAE 382 is selected as the typical locality.

*Geographic Distribution:* Two localities, near Jiancaogou (AAE 382) and Jiadanwan (AAE 365), near Zunyi, Guizhou Province.

*Environment:* The fossils occur in a very calcareous, fine- to medium-grained siltstone-sandstone that includes much comminuted, coarse-sand-sized organic debris (including pelmatozoan columnals). The bivalved shells are thoroughly disarticulated and sorted. Broken shells make up a small percentage of the total. The fossil bed is about 30 cm thick (Zhang and others, 1964), is underlain by the Wufeng Formation and overlain by the Lungmachi Formation. The population density of the specimens approaches that of a shell bed. The brachiopods commonly are found in the convex-side-up position. The available material is highly weathered, so that its original color is not known. The much higher diversity of this fauna, which includes abundant trilobites and other nonbrachiopod taxa, suggests a more normal environment than that represented by the *Aphanomena-Hirnantia* Community. Rong concluded that the *Aphanomena-Hirnantia-Dalmanitina* Community occupied a more shoreward position than did the *Aphanomena-Hirnantia* Community at the type locality, as well as many of the other localities. The overall aspect of the fauna suggests a relatively normal, although not turbulent water, Benthic Assemblage 3 fauna. The larger size of the *Hirnantia* specimens is consistent with a close relationship to Benthic Assemblage 2 in which specimens of a taxon tend to be of larger size (Boucot, 1975). The thoroughly disarticulated nature of all elements (trilobite, brachiopod, pelmatozoan, bivalve) is consistent with moderate current activity, as is the high concentration of calcareous organic debris in the Hirnantian strata at the two localities. The high-diversity fauna suggests that the factors responsible for the low diversity of the allied *Aphanomena-Hirnantia* Community were not operative here.

This fauna belongs to the same biogeographic unit as do other South China Hirnantian faunas.

*Community Group:* Most Ordovician communities cannot yet be assigned to a community group.

*Atrypa–Xystostrophia* Community (Table 4)

*Name:* This community is named here.

*Age:* Eifelian (see Anderson and others, 1969).

*Composition:* A high-diversity community, including many

brachiopods and corals, with the latter being chiefly solitary rugose corals.

**Typical Locality:** Padaukpin, Burma.

**Geographic Distribution:** Presently known only from the Shan States of Burma at Padaukpin, but biogeographic considerations (see Wang and others, 1984) suggest that this community is not to be expected in the marine Eifelian of South China and adjacent parts of North Vietnam. The Rhenish-Bohemian Region aspect of this fauna (Wang and others, 1984) was also suggested by Struve (1982, Fig. 1).

**Environment:** The fossils are chiefly articulated specimens in a medium-bedded, calcareous mudstone and thin-bedded, bioclastic limestone. This suggests a moderately quiet-water, normal, Benthic Assemblage 3 marine environment.

**Community Group:** *Striispirifer.*

**Atrypoidea Community** (Figs. 11–12, Table 5)

**Name:** This community is recognized in the sense of Boucot (1975), but with the generic change from *Atrypella* to *Atrypoidea* made necessary by Copper's (1977b) taxonomic revision; it is a low-diversity unit dominated by *Atrypoidea.*

**Composition:** Large *Atrypoidea* with a few rugose corals occur in Inner Mongolia (Su, 1976). Chiefly *Atrypoidea* Yunnan (Table 5).

**Age:** Ludlovian-Pridolian (Mu and others, 1986; Li and others, 1984). Ludlovian (Wang and others, 1980).

**Typical Locality:** None designated.

**Geographic Location:** Gashaomiao section, Daerhanmaomingan, Inner Mongolia (Su, 1976), Qujing, Yunnan.

**Environment:** About two-thirds to three-quarters of the Inner Mongolian specimens are disarticulated, which is unusual for the *Atrypoidea* Community. The material occurs in gray, argillaceous, calcarenitic limestone. The low-diversity, high-dominance nature of the collection is consistent with the quiet-water, restrictive conditions common to the *Atrypoidea* Community in many parts of the Uralian-Cordilleran Region during the Upper Silurian. The disarticulated nature of the shells from Inner Mongolia, combined with the calcarenitic nature of the matrix and the size-sorted valves, suggests postmortem sorting. A Benthic Assemblage 2 or shallow 3 assignment is reasonable after comparison with this community in other parts of the world. Inspection of the Gashaomiao section in 1983, under the direction of Dr. Li Wenguo, showed that the actual *Atrypoidea* bed is about 0.33 m thick, and occurs immediately beneath unfossiliferous strata of nonmarine, Old Red Sandstone aspect. In fact, a few valves of *Atrypoidea* were noted in the top of the *Atrypoidea* bed actually imbedded in red siltstone. A few spiriferids were also noted with the corals and *Atrypoidea* (pedicle valves) specimens. Possibly the shells in this bed were transported shoreward from Benthic Assemblage 2 into 1, or perhaps part of the overlying, unfossiliferous material of nonmarine aspect actually belongs to Benthic Assemblage 1. In any event, it is clear that for the first time the Gashaomiao section *Atrypoidea* bed provides unequivocal evi-

**TABLE 4. ATRYPA-XYSTOSTROPHIA COMMUNITY***

| Name | Total No. Specimens | % |
|---|---|---|
| Atrypa | 200+ | 20 |
| Xystostrophia | 87 | 9 |
| Aulacella | 50 | 5 |
| Schizophoria | 35 | 3 |
| Mystrophora | 3 | – |
| Kayserella | 14 | 1 |
| Sieberella | 14 | 1 |
| Leptaena | 6 | 1 |
| Mesoleptostrophia (their Leptostrophia) | 3 | – |
| Leptodontella | 16 | 2 |
| Mesodouvillina | 50 | 5 |
| Talaeoshaleria | 38 | 4 |
| Radiomena | 3 | – |
| Devonaria | 22 | 2 |
| Productella | 2 | – |
| Uncinulus | 79 | 8 |
| Markitotoechia | 9 | 1 |
| Schnurella | 1 | – |
| Septalaria? | 1 | – |
| Athyris | 20 | 2 |
| Nucleospira | 52 | 5 |
| Plectospira | 32 | 3 |
| Plectospira | 3 | – |
| Invertrypa | 51 | 5 |
| Desquamatia | 33 | 3 |
| Indospirifer | 17 | 2 |
| Alatiformia? | 3 | – |
| Reticulariopsis | 23 | 2 |
| Emanuella | 44 | 4 |
| Cyrtina | 69 | 8 |
| Cimicinoides | 30 | 3 |
| Merista | 11 | 1 |
| Total | 1,021+ | |

*From same locality and collection as Anderson, M. M. and others, 1969.

dence of the very shallow-water, nearshore position of some *Atrypoidea* Community occurrences.

**Community Group:** *Striispirifer.*

**Atrypoidea–Protathyrisina uniplicata–Striispirifer Community** (Fig. 11; Table 6).

**Name:** This community is named here.

**Composition:** *Striispirifer* and *Protathyrisina uniplicata* are dominant, associated with a variable percentage of *Atrypoidea* (Table 5).

**Age:** Ludlovian (Wang and others, 1980).

**Typical Locality:** Qujing, eastern Yunnan Province.

**Geographic Distribution:** Eastern Yunnan only, in the Kuanti Formation.

**Environment:** The shells occur as well-sorted, articulated individuals in a gray, very calcareous mudstone. The high degree of articulation, as well as the low diversity, indicates a quiet-water environment, with some restrictive factors. An inner Benthic Assemblage 3 or outer 2 (Jones and Rong, 1982) environment is indicated by the occurrence of *Striispirifer.*

**Community Group:** *Striispirifer.*

TABLE 5. ATRYPOIDEA COMMUNITY

| Name | Articulated Shell | Pedicle Valve | Brachial Valve | Either Valve | Total No. Specimens | % |
|---|---|---|---|---|---|---|
| **Locality ADH 28, Kuanti Formation** | | | | | | |
| Atrypoidea qujingensis | 82 | | | | 82 | 22 |
| Atrypoidea dorsoconvexa | 15 | | | | 15 | 4 |
| Atrypoidea sp. | 215 | 4 | | | 219 | 59 |
| Protathyrisina uniplicata | 1 | | | | 1 | |
| Protathyrisina sp. | 39 | 2 | | 5 | 46 | 12 |
| Striispirifer yunnanensis | 3 | 4 | | | 7 | 2 |
| Total | | | | | 370 | |

TABLE 6. ATRYPOIDEA-PROTATHYRISINA UNIPLICATA-STRIISPIRIFER COMMUNITY

| Name | Articulated Shell | Pedicle Valve | Brachial Valve | Either Valve | Total No. Specimens | % |
|---|---|---|---|---|---|---|
| **Locality ADH 25, Kuanti Formation** | | | | | | |
| Atrypoidea qujingensis | 5 | 1 | | | 6 | 4 |
| Protathyrisina uniplicata | 24 | 5 | | 3 | 32 | 22 |
| Striispirifer yunnanensis | 40 | 38 | 20 | 10 | 108 | 74 |
| Total | | | | | 146 | |
| **Locality ADH 26, Kuanti Formation** | | | | | | |
| Atrypoidea qujingensis | 1 | 2 | | | 3 | 1 |
| Protathyrisina uniplicata | 107 | 35 | 52 | 10 | 204 | 55 |
| Striispirifer yunnanensis | 41 | 82 | 29 | 9 | 161 | 44 |
| Total | | | | | 368 | |

### *Beitaia–Eospirifer* Community (Fig. 9)

*Name:* This community is named here.
*Composition:* This is a low-diversity community, with specimens of the nalivkiniid atrypacean *Beitaia* occurring about three times as frequently as *Eospirifer,* and a few specimens of *Kritorhynchia.*
*Age:* The community is of Middle Llandoverian age (Rong and Yang, 1978, 1981).
*Typical Locality:* Leijiatun, in the lowest bed of the Xiangshuyuan Formation.
*Geographic Distribution:* The community is well developed at Leijiatun, Shiqian County, northeastern Guizhou Province.
*Environment:* This community consists of abundant, articulated specimens preserved in a highly calcareous gray mudstone. The beds yielding the community overlie the Lungmachi Formation, with its graptolitic fauna; they are overlain by a coral-brachiopod–rich sequence of the Xiangshuyuan Formation. The *Beitaia-Eospirifer* Community occurs in a bed having a thickness of about 2 m. The shells are well sorted.

The environment represented by this community was a quiet one, as suggested by the low diversity and abundance of articulated brachiopods. The position between a Benthic Assemblage 3, coral-rich unit and a low-diversity, graptolite-rich unit makes a Benthic Assemblage 3 assignment reasonable, as does the general similarity to the *Dubaria* and *Dayia* Communities. The population density of the shells in the community is fairly high. It may be that a low-oxygen, relatively quiet-water environment provided the restrictive community condition.

Biogeographically this community is of interest because of the abundance of the endemic South China Llandoverian genus *Beitaia.*
*Community Group: Striispirifer.*

### Bivalve Community (Fig. 14)

*Name:* See Boucot (1975) for a brief discussion under "Pelecypod Community."
*Age:* The upper part of the Lianhuashan Formation that contains Bivalve Community material is of Siegenian age because it has a gradational relation with the overlying, well-dated Nahkaoling Formation of Siegenian age (see "*Orientospirifer-Sinochonetes* Community"), and also because the vertebrates it contains are consistent with a Siegenian age. Pan and others (1978) listed *Yunnanolepis* sp., *Lianhuashanolepis liukingensis, Orientolepis neokwangsiensis, Asiaspis expansa, Asiacanthus suni, A. kaoi*

from the Lianghuashan Formation. These vertebrate taxa also occur in the Xitun Formation of the Qujing region, eastern Yunnan.

*Composition:* Several bivalve taxa and a linguloid. In the Nahkaoling Formation these shells are very scattered.

*Typical Locality:* None selected. The term Bivalve Community will certainly be properly changed into generic and specific level communities assigned to appropriately defined community groups dominated by different bivalve taxa. At that time it will be appropriate to select type localities for the different bivalve-dominated communities and also for the vertebrate-dominated communities.

*Geographical Distribution:* Liujing area, Hengxian County, South Guangxi.

*Environment:* As discussed by Boucot (1975), the Bivalve Community is commonly a Benthic Assemblage 1 unit. The presence of inarticulate brachiopods in low frequency, as well as vertebrates, is consistent with this conclusion.

*Community Group:* Until the bivalves in these bivalve-dominated communities are studied, it would be premature to devise a community group nomenclature.

### *Borealis* Community (Fig. 9)

*Name:* The term *Borealis* Community is used here in Boucot's sense (1975).

*Composition: Borealis* only. The sample consists of about 20 specimens that were collected by Rong and Yang from an approximately 50-cm thick argillaceous limestone bed packed with specimens of *Borealis* to the exclusion of other fossils.

*Age:* Late Middle Llandoverian (Rong and Yang, 1981).

*Typical Locality:* See Rubel (1970) for discussion of the *Borealis* Bank localities in Estonia, which are designated the typical localities.

*Geographic Distribution:* This community is known in China only from four North Guizhou localities, but it is abundant in Europe (see Rubel, 1970) and some other parts of the northern Old World. The Chinese localities are at Shiqian, Wuchuan, and Sinan County, northern Guizhou Province.

*Environment:* The fossils are mostly articulated and dispersed in argillaceous, gray limestone. Their occurrence immediately above graptolitic shales, containing *M. gregarius* Zone graptolites, is consistent with a Benthic Assemblage 3 position. The articulated condition of the shells at three localities is unusual for pentameroids, and indicates relatively quiet-water conditions. At the Wuchuan locality the shells are thoroughly disarticulated as they are at many European localities (the general case for pentameroids).

The presence of *Borealis* is biogeographically interesting because the genus is absent on the North American Platform, where its place is taken by *Virgiana,* but is present in the Baltic region and on the Siberian Platform where both genera occur.

The quiet-water Benthic Assemblage 3 *Borealis* Community occurrences, immediately above the graptolitic community, occa-

sion no difficulty. Graptolitic communities of the Llandoverian may extend from Benthic Assemblage 2 through 6 (Berry and Boucot, 1972), *but* their taxic diversity is very low in 2 (about two genera), increasing to high in Benthic Assemblage 6. Lin Yao-kun (oral communication, 1980) reports that the graptolite fauna found immediately beneath the *Borealis* Community is of moderately low diversity (*Pristiograptus, Monograptus, Pseudoclimacograptus, Climacograptus*) rather than the much higher diversity expected in the mid-Llandoverian deeper water facies.

*Community Group:* Virgianinae. Boucot (1975, p. 262) referred to "Virgianinid Communities" that include many of the virgianid genera, but he did not formally designate a Virgianidae Community Group consisting of high-dominance, low-diversity, almost monospecific aggregations of abundant, varied *virgianid* genera such as *Borealis, Eoconchidium, Tscherskidium, Holorhynchus, Nondia, Platymerella, Paraconchidium, Pleurodium, Pseudoconchidium, Virgiana,* and *Virgianella.*

### *Brevilamnulella* Community (Fig. 9)

*Name:* The term *Brevilamnulella* Community is used here in the sense discussed under *"Stricklandia"-Merciella* Community, for a unit with abundant *Brevilamnulella* and lacking *Cyrtia.*

*Composition:* Abundant *Brevilamnulella* together with *Zygospiraella* and *Beitaia.*

*Age:* Late Middle Llandoverian (Rong and Yang, 1981).

*Typical Locality:* Heshui, Yinjiang, northeastern Guizhou.

*Geographic Distribution:* Heshui, Yinjiang County, northeastern Guizhou.

*Environment:* The presence of abundant articulated specimens of *Brevilamnulella* associated with only two other brachiopod genera in a gray, calcareous mudstone is consistent with a restrictive Benthic Assemblage 5, fairly quiet-water environment.

*Community Group:* Boucot (1975, p. 262) used the term "Undivided Communities" for units of moderate to high diversity containing a moderate to high abundance of *Cyrtia* and/or *Brevilamnulella* lacking abundant or even moderate numbers of *Dicoelosia* and/or *Skenidioides.* In the discussion of the Community Group assignment of the *"Stricklandia"-Merciella* Community, we recommend that Boucot's (1975, p. 262) "Undivided Communities" be termed the *Brevilamnulella-Cyrtia* Community. We further suggest that Benthic Assemblage 5 units dominated by *Brevilamnulella,* or the descendent taxon *Clorinda,* be termed the *Brevilamnulella* and *Clorinda* Communities, respectively. All these units belong to the *Dicoelosia-Skenidioides* Community Group. Units dominated by *Cyrtia* can be termed the *Cyrtia* Community, with a soft substrate being indicated, as at Golden Grove Park in the Welsh Borderland Wenlockian. A "soft" substrate is probable because the broad pedicle valve interarea of *Cyrtia* functionally is consistent with support on a soft substrate.

### *Buchiola-Reticulariopsis* Community (Figs. 16 and 17)

*Name:* This comunity is named here.

*Age:* Late Emsian and Early Eifelian (see Wang Cheng-yuan and others, 1979) with abundant *Nowakia holynensis, N. richteri,* and abundant goniatites, from the Nandan Formation.

*Composition:* Xu Han-kui (oral communication, 1980) stated that *Buchiola* is very abundant and that (in descending order) the brachiopods *Reticulariopsis, Perichonetes,* and *Costanoplia* are most abundant; however, the brachiopods are very scattered. Two other notanopliid genera, *Paraplicanoplia* and *Tangxiangia,* are rare, as are trilobites. See Xu Han-kui (1979, Table 1) for a complete faunal list.

*Typical Locality:* Tangxiang-Nabiao, Luofu, Nandan County, Northern Guangxi.

*Geographic Distribution:* Only the typical locality, although a similar community may occur in Lingshan County, southern Guangxi.

*Environment:* Xu Han-kui reported (oral communication, 1980) that the brachiopods occur as very scattered, chiefly disarticulated specimens, all small. The beds (about 0.5 cm thick, on the average; Xu Han-kui, oral communication, 1980) consist of dark-colored, thin-bedded mudstone. The *Nowakia* and goniatites are pelagic, and the trilobites are probably epibenthic. We interpret these strata as Benthic Assemblage 4–5, deposited in a relatively quiet-water, lower oxygen environment, as indicated by the thin-bedded, nonbioturbated strata. This community has much in common with the *Maoristrophia* Community of eastern Australia (Boucot, 1975).

*Community Group: Dicoelosia-Skenidioides.*

### *Coelospirella* Community (Fig. 15)

*Name:* This community is named here.

*Composition:* Su Yang-zheng reported (written communication, 1980) dominant *Coelospirella* with less abundant specimens including *Megastrophia,* "*Chonetes,*" *Cyrtina, Spinella, Brachyspirifer, Paraspirifer, Merista, Borealirhynchia, Rhytistrophia,* and *Howellella,* as well as other genera from sample to sample.

*Age:* Late Siegenian to Emsian (Su Yang-zheng, written communication, 1980), lowest part, second part, and fourth part of five faunal divisions of the Obotanhondi Formation.

*Typical Locality:* None designated.

*Geographic Distribution:* Dong Ujimqin Qi (County), Neimongol. Similar coelospirinid-rich Siegenian and Emsian units occur in the Eastern Americas Realm of eastern North America. This unit is not the same as the *Coelospira-Pacificocoelia* ("*Coelospira-Leptocoelia*" Community of Boucot, 1975) Community of the Eastern Americas Realm, which occurs in a very quiet-water, thin-bedded, low population density environment. But, similar high-dominance, low-diversity Silurian occurrences of coelospirinids occur in units such as the Kenneth Limestone of Indiana.

*Environment:* The disarticulated nature of the shells, and the sand-sized matrix is consistent with normal current activity, as is the presence of less common *Coelospirella* in the *Sinostrophia-Discomyorthis* high-diversity, low-dominance community. We assign the *Coelospirella* Community to Benthic Assemblage 3 because of its intimate association with the *Sinostrophia-*

*Discomyorthis* Community.

*Community Group: Striispirifer.*

### *Cryptatrypa–Strophochonetes* Community (Fig. 16)

*Name:* This community is named here.

*Age:* Late Emsian (Wang Cheng-yuan and others, 1979) as indicated by the associated *Nowakia cancellata* and goniatites in this Nandan Formation community.

*Composition:* See Xu Han-kui (1979, Table 1) for a complete faunal list. Among the brachiopods *Cryptatrypa, Strophochonetes* and *Muriferella,* in descending order, are the most abundant. The notanopliid *Costanoplia* is rare. Trilobites are uncommon, while *Nowakia* and goniatites are abundant.

*Typical Locality:* Tangxiang-Nabiao Section, Luofu, Nandan County, northern Guangxi. A similar unit may occur in Lingshan County, southern Guangxi.

*Geographic Distribution:* Same as above.

*Environment:* Xu Han-kui reported (oral communication, 1980) that the brachiopods occur as very scattered, chiefly small, disarticulated specimens. The beds consist of dark-colored, thin-bedded (about 0.5 cm thick, on the average; Dr. Xu, oral communication, 1980) mudstone. The *Nowakia* and goniatites are pelagic, and the trilobites probably epibenthic. We interpret these strata as Benthic Assemblage 4-5, deposited in a relatively quiet-water, lower oxygen environment, as indicated by the thin-bedded, nonbioturbated strata. This community has much in common with the *Maoristrophia* Community of eastern Australia, as well as some of the other deeper water Devonian communities containing scattered plectodontids and notanopliids.

*Community Group: Dicoelosia-Skenidioides.*

### *Dicoelosia–Skenidioides* Community Group High Diversity Community (Fig. 12)

*Name:* We use the term in the sense of Boucot (1975) for a high-diversity community in the Benthic Assemblage 4-5 range.

*Composition:* This high-diversity community has been recognized at two places in northeastern China. At Erdaogou, Liu and Huang (1977) found the following in decreasing abundance: *Delthyris, Leptaena, Lissatrypa, Nucleospira,* and *Skenidioides.* They found lesser numbers of *Isorthis, Atrypa, Dicoelosia, Coolinia,* and *Protochonetes,* together with rugose corals, tabulates, and trilobites. They also stated that the lower part of the same stratigraphic unit at Erdaogou contains two assemblages: a lower assemblage, in which *Delthyris, Leptaena, Nucleospira,* and *Skenidioides* are present in decreasing abundance, together with abundant *Mucophyllum, Rhizophyllum,* and tabulates; and an upper assemblage, which contains, in decreasing abundance, *Isorthis, Dicoelosia,* and *Atrypa,* with abundant *Spongophyllum* and *Disphyllum.*

In the Lesser Khingan Mountains, Xue and others (1980) reported the following from the lower Woduhe Formation below a *Tuvaella gigantea* Community: *Craniops, Dolerorthis* spp., *Dico-*

TABLE 7. <u>EOSCHIZOPHORIA HESTA-PROTATHYRIS XUNGMIAOENSIS</u> COMMUNITY

| Name | Articulated Shell | Pedicle Valve | Brachial Valve | Either Valve | Total No. Specimens | % |
|---|---|---|---|---|---|---|
| Locality ADH 59, Miaokao Formation | | | | | | |
| <u>Eoschizophoria</u> <u>hesta</u> | 313 | 39 | | 28 | 380 | 56 |
| <u>Protathyris</u> <u>xungmiaoensis</u> | 268 | 14 | | | 282 | 42 |
| <u>P.</u> <u>planosulcata</u> | 6 | | | | 6 | 1 |
| <u>Howellella</u> <u>tingi</u> | 1 | | 1 | | 2 | |
| <u>Protathyrisina</u> <u>uniplicata</u> | 1 | | | | 1 | |
| <u>P.</u> <u>minor</u> | 2 | | | | 2 | |
| <u>P.</u> <u>plicata</u> | 1 | | | | 1 | |
| Total | | | | | 674 | |
| Locality ADH 58-59, Miaokao Formation | | | | | | |
| <u>Eoschizophoria</u> <u>hesta</u> | 58 | 24 | | | 82 | 35 |
| <u>Protathyris</u> <u>xungmiaoensis</u> | 40 | 3 | | | 43 | 18 |
| <u>P.</u> <u>planosulcata</u> | 4 | | | | 4 | 2 |
| <u>Protathyrisina</u> <u>uniplicata</u> | 22 | | | | 22 | 9 |
| <u>P.</u> <u>minor</u> | 61 | 1 | | | 62 | 26 |
| <u>P.</u> <u>plicata</u> | 10 | | | | 10 | 4 |
| <u>Aesopomum</u> <u>delicatum</u> | | 1 | | | 1 | |
| <u>Howellella</u> <u>tingi</u> | 3 | | | | 3 | 1 |
| <u>Spirinella</u> sp. | 7 | 3 | | | 10 | 4 |
| Total | | | | | 237 | |

*elosia, Pentlandina, Leptaena, Atrypa, "Coelospira," Meristina, Cyrtia, Eospirifer, Janius, Isorthis, Plectatrypa, "Camarotoechia," Dalmanophyllum, Tryplasma, Microspongia.* This association appears to belong to the *Striispirifer* Community Group.

*Age:* The community is present in the lower part of the Erdaogou Formation, which may be of Pridolian age. The presence of *Protathyrisina* in the Erdaogou (Liu and Huang, 1977) indicates an age of Pridolian or older, whereas the presence of the notanopliid *Septoparmella* ("*Metaplasia*" of Liu and Huang, 1977) in the overlying upper member suggests a Devonian (i.e., Gedinnian) age assignment.

The Lower Woduhe Formation contains *Janius* and "*Coelospira,*" indicating a position no older than the Late Wenlockian; *Pentlandina,* suggesting an age no higher than the Wenlockian; and *Cyrtia,* indicating a pre-Pridolian age.

*Typical Locality:* None designated.

*Geographic Distribution:* Erdaogou, Yongji County, Jilin Province, and Woduhe, Lesser Khingan Mountains.

*Environment:* As the Erdaogou Formation lies above a Benthic Assemblage 3 unit, and underlies a Benthic Assemblage 4-5 unit, this community unit may represent Benthic Assemblage 4, and a relatively normal, quiet-water environment. Liu and Huang (1977) reported calcareous shale with limestone lenses in the lower part of the Erdaogou Formation. Their illustrations indicate disarticulated valves, occurring as casts and molds, in the clastic rocks. Presumably, current activity was sufficient to disarticulate the shells.

The Woduhe occurrence is below a Benthic Assemblage 3 *Tuvaella gigantea* Community, which suggests an upward-shallowing sequence.

*Community Group: Striispirifer.*

### *Eoschizophoria hesta–Protathyris xungmiaoensis* Community (Fig. 11, Table 7)

*Name:* This community is named here.

*Composition:* Table 7 indicates the medium diversity of this unit, and its dominance by *Eoschizophoria* and *Protathyris,* as well as a significant admixture of *Protathyrisina* in one sample. This last sample may represent a community boundary mixture with the *Protathyrisina* Community.

*Age:* Late Ludlovian–Early Pridolian (Wang and others, 1980; Rong and Yang, 1980).

*Typical Locality:* Qujing, eastern Yunnan Province.

*Geographic Distribution:* Eastern Yunnan only in the Miaokao Formation.

*Environment:* The largely articulated condition of the material, and its occurrence in very calcareous mudstone, indicates a relatively quiet-water environment. The community bears many similarities to both the *Protathyris* and *Protathyrisina* Communities, with which it is closely associated.

*Community Group: Striispirifer.*

### *Eospirigerina* Community (Fig. 8)

*Name:* This community is named here.

*Composition: Eospirigerina* only. The sample consists of approximately 20 specimens.

*Age:* Rong (1979) summarized the reasons for assigning the Wulipo Bed to the Lower Llandoverian.

*Typical Locality:* Wulipo, Meitan, northern Guizhou, in the upper part of the Wulipo Bed.

*Geographic Distribution:* In addition to the Wulipo occur-

TABLE 8. EOSPIRIGERINA-HINDELLA COMMUNITY

| Name | Articulated Shell | Pedicle Valve | Brachial Valve | Either Valve | Total No. Specimens | % |
|------|-------------------|---------------|----------------|--------------|---------------------|---|
| **Locality AAE 506** | | | | | | |
| Hindella | 2 | 63 | 14 | 5 | 84 | 12 |
| Eospirigerina | | 57 | 26 | 473 | 556 | 81 |
| Orthids | | 2 | | 2 | 4 | 1 |
| Eospirifer | 2 | 12 | 10 | | 24 | 4 |
| Strophomenid | 1 | 2 | | 5 | 8 | 1 |
| Rostricellula | | | 2 | 3 | 5 | 1 |
| Dalmanella | 1 | 1 | | | 2 | |
| Total | | | | | 683 | |

rence, this community has been recognized in the Lake District, northern England (Marr and Nicholson, 1888, described a fauna that belongs to this community). Communities dominated by the descendant genus *Spirigerina* are present in the younger Silurian and even the Lower Devonian in other parts of the world.

*Environment:* The specimens occur in a very calcareous, gray mudstone in a largely articulated, scattered condition. They do not appear to be in life position. The presence of a high percentage of articulated shells suggests a fairly quiet-water environment, but the absence of life-position orientation suggests a certain level of water movement or bioturbation. The very low-diversity, high-dominance assemblage indicates a very specialized environment, with relatively little water movement. A Benthic Assemblage 3 location is reasonable, in view of the underlying *Eospirigerina-Hindella* Community in the lower part of the Wulipo Bed and the nature of the graptolitic fauna of the overlying Lungmachi Formation. One possible explanation of the low diversity might be that low-oxygen conditions of the kind envisioned for Lungmachi Formation bottoms might have been little different from those present during *Eospirigerina* bed time. Such an interpretation would be consistent with the cosmopolitan and relatively eurytopic distribution of *Eospirigerina* and *Spirigerina* (indicated by their presence in many Benthic Assemblage 3-5 communities).

*Community Group: Striispirifer.*

### *Eospirigerina–Hindella* Community (Fig. 8; Table 8)

*Name:* This community is named here.
*Composition:* Table 8 lists material from a locality showing the low- to medium-diversity brachiopod fauna characterized by a dominance of *Eospirigerina* and an abundance of *Hindella*.
*Age:* Rong (1979) summarized the reasons for assigning the Wulipo Bed, which contains this community, to the Lower Llandoverian.
*Typical Locality:* The Wulipo occurrence (Locality AAE 506) is selected as the typical locality.
*Geographic Distribution:* The community occurs at Wulipo, Meitan County, Jiadanwan and Jiancaogou, Donggongsi, Zunyi,

northern Guizhou Province, and at Leijiatun, Shiqian County, northeastern Guizhou Province (Rong, 1979).
*Environment:* The fossil bed (Locality AAE 506) is 0.4 m thick, and consists of gray, calcareous mudstone. The shells are thoroughly disarticulated, but mostly unbroken. Most of the valves rest in a convex side upward position. The location above the unconformably underlying Caradocian Pagoda Limestone, and below the *Eospirigerina* bed and the Lungmachi Formation, is consistent with a shallow-water interpretation. The relatively low diversity of the fauna, as well as the abundance of *Eospirigerina* and *Hindella* (itself very similar to *Cryptothyrella*), suggests a very shallow Benthic Assemblage 3 position. Bottom conditions were probably moderately turbulent, as indicated by the disarticulated condition of most of the brachiopod shells. Dominance of *Cryptothyrella* or *Hindella* would indicate a Benthic Assemblage 2 position.

Biogeographically, this community is assigned to the North Silurian Realm. It is definitely extra-Malvinokaffric Realm in nature.
*Community Group: Striispirifer.*

### Graptolitic Community (Pelagic Community) (Figs. 8, 9 and 10)

*Name:* We do not formally assign names to this unit for either the Ashgillian Wufeng Formation or for the Llandoverian Lungmachi Formation, but regard it as the same basic unit discussed briefly in Boucot (1975; "Pelagic Community"). These graptolitic communities are essentially the well-known graptolitic facies of other workers.
*Composition:* The megafauna consists almost exclusively of abundant, flattened graptolites. No attempt has been made here to divide the Llandoverian Lungmachi Formation graptolitic associations into communities, but Chen Xu (*in* Mu and others, 1986) has done so.
*Typical Locality:* None is designated, as this study emphasizes the shelly, brachiopod-rich, rather than graptolitic, communities.
*Environment:* Both the Wufeng and Lungmachi Formations are

characterized by thinly bedded, black shales, except for the upper part of the Lungmachi, which has an olive-drab color. The beds are characteristically platy and nonlaminated. Bentonites may be present in the Lungmachi as thin "clay" layers. The dark color is probably due to both abundant free carbon and fine-grained pyrite. The thinly bedded (averaging about 5 to 15 mm thick), nonbioturbated beds suggest a very quiet-water, low-oxygen environment; this is also consistent with the almost total absence of benthic macrofauna. These two formations are assigned a Benthic Assemblage 3-5 position because of their paleogeographic location seaward of the contemporary shelly faunas and because of their graptolite faunas (Chen Xu, *in* Mu and others, 1986). Their lithology and great lateral extent have much in common with similar, contemporary units in other parts of the world: the Cape Phillips, Road River, and Roberts Mountains Formations of North America; and a similar Silurian unit that covers much of North Africa and Europe west of the Russian Platform. Similar blankets of black shale yielding a pelagic fauna are represented by the Upper Devonian–Lower Mississippian Chattanooga Shale of North America; this is also assigned approximately to Benthic Assemblage 6.

Ordinarily, the Graptolitic Community is thought of as a strictly Benthic Assemblage 6, shelf margin region, unit. Berry and Boucot (1972) further discussed this matter of the Benthic Assemblage range of the graptolitic facies, and show that a Benthic Assemblage range of 2-5 is involved. Chen Xu (in Mu and others, 1986) has treated this question in some detail for the Chinese Silurian, and shows that a 2-5 range is reasonable from locality to locality.

*Community Group:* Until the community ecology of the graptolites becomes better understood, it would be premature to assign community groups.

## "*Euryspirifer*" *qijianensis-Kwangsia* Community (Fig. 16)

*Name:* This community is named here.
*Age:* Late Emsian (Wang Cheng-yuan and others, 1979), based on conodonts.
*Composition:* Abundant "*Euryspirifer,*" and *Athyrisina,* together with the ostracode *Paramoelleritia,* together with less abundant *Lazutkinia, Indospirifer, Xenostrophia?,* and "*Uncinulus?*"
*Typical Locality:* Dingshanling, Xiangzhou County, Guangxi, in the Dingshanling Member of the "Sipia Formation."
*Geographic Distribution:* Guangxi, Guizhou, N. Sichuan (Wang and Zhu, 1979).
*Environment:* Most of the shells occur as articulated individuals in a very calcareous shale matrix, which suggests a relatively normal, quiet-water environment. The small size of the available sample (about 100 shells) does not lend itself to a rigorous treatment, but this community appears to be a typical Benthic Assemblage 3, normal diversity unit.
*Community Group: Striispirifer.*

## *Fallaxispirifer–Discomyorthis* Community (Fig. 15)

*Name:* This community is named here.
*Composition:* Su Yang-zheng (written communication, 1980) reported abundant *Fallaxispirifer,* less abundant *Discomyorthis,* still less abundant *Tridensilis* and *Spinatrypa,* and even less abundant *Rhytistrophia, Borealirhynchia,* and *Coelospirella,* together with rare *Maoristrophia?, Wyella, Pseudochonetes, Cyrtina, Schizophoria, Leptaenopyxis, Howellella, Merista, Megakozlowskiella, Paraspirifer,* and *Sinostrophia.*
*Age:* Late Siegenian to Emsian, from the highest (fifth) horizon of the Obotanhondi Formation (Su Yang-zheng, written communication, 1980).
*Typical Locality:* None designated.
*Geographic Distribution:* Only at Dong Ujimqin Qi (County), Neimongol, at the present time, although *Fallaxispirifer* is known from the Central Asian part of the Soviet Union to Heilongjiang in the People's Republic of China.
*Environment:* The generally disarticulated nature of the fauna, the sandy matrix, the high diversity, and the relatively low dominance, all suggest a Benthic Assemblage 3, relatively normal marine environment. The differences between this unit and the *Sinostrophia-Discomyorthis* Community are concerned more with abundance than overall taxic composition.
*Community Group: Striispirifer.*

## Gypidulid–Uncinulid Community (Fig. 12)

*Name:* This community is named here.
*Composition:* Generically indeterminate gypidulinid and uncinulid.
*Age:* Ludlovian-Pridolian interval (see Mu and others, 1986).
*Typical Locality:* Darehan Maomingan Lianheqi, Inner Mongolia.
*Geographic Distribution:* Same as above.
*Environment:* Scattered disarticulated valves of the two taxa are preserved in a coarse-grained, micaceous, feldspathic sandstone occurring not far above crystalline basement. This suggests the presence of a turbulent environment not too far from shoreline, as indicated by the coarse sediment, which is consistent with the presence of the gypidulids and the low diversity, i.e., a restrictive environment. A Benthic Assemblage 3 assignment is indicated by the presence of a gypidulid.

Despite the poor quality of the specimens, this material is of sufficient ecologic significance to warrant being singled out for attention.
*Community Group:* Gypidulinidae.

## *Harpidium* Community (Fig. 10)

*Name:* The *Harpidium* Community is another of the many Pentameridae Community Group Communities.
*Composition:* Consists of almost 100 percent *Harpidium,* com-

monly in a disarticulated condition at many non-Chinese localities. Two of the South China examples consist of articulated shells brought some years ago to Wang Yu by a geologist since deceased. As pointed out by Boucot and Johnson (1979), articulated Pentameridae Group occurrences are exceptional; they are consistent with the thesis that turbulent conditions were necessary for larval settlement and success, although not necessarily for adult growth, and that turbulent water conditions present for larval settlement commonly persist through the adult stage as well.

*Age:* Shihniulan Formation of early Late Llandoverian age (Mu and others, 1986).

*Typical Locality:* None designated.

*Geographical Distribution:* Two localities in northern Guizhou, one in Wuchuan County and the other in Tongzi County. These shells were originally described by Wang (1955) as *Pentamerus muchuanensis,* and are reassigned by us to *Harpidium (Isovella)* because of their nontrilobate, slightly bisulcate form, coupled with a relatively deep pedicle valve, and a pedicle valve median septum reaching only about halfway to the anterior margin (Rong and Yang, 1981).

*Environment:* At two localities the shells occur in argillaceous limestone beds about 1 to 2 m thick, in an articulated condition. An articulated condition is unusual for pentamerid occurrences, although not unknown (Boucot and Johnson, 1979). The articulated condition indicated that the larvae settled on a firm mud or sand bottom over which there were reasonably active, turbulent conditions. Following the initial settlement, conditions became less turbulent, which allowed the initial set to grow to adulthood, but prevented future larval settlement. We conclude that the ecologically critical stage for Pentameridae is the larval settling state requiring rough water, rather than the adult stage that can develop in either rough- or quiet-water conditions. A Benthic Assemblage 3 assignment is made because this is where other Pentameridae Community Group shells invariably occur. The Chinese occurrences are in relatively quiet-water conditions for the adults, but are shown on the community framework (Fig. 10) in the rough-water position normally common to this Community Group because of the larval considerations discussed above; that is, we conclude that rough-water conditions were necessary for larval settlement and survival, although not necessarily for adult success.

*Community Group:* Pentameridae.

### Harpidium–Stricklandia–Merciella Community (Fig. 10)

*Name:* This community is named here.

*Composition:* This is a moderately high-diversity unit characterized by a reasonable abundance of *Harpidium, Stricklandia,* and *Merciella,* plus *Isorthis, Zygospiraella, Brevilamnulella, Eospirifer, Dinobolus, Mesoleptostrophia, Katastrophomena, Pleurodium, Lissatrypa,* and *Aegiria,* together with a relatively rich shelly fauna of rugose corals, trilobites, stromatoporoids, and graptolites, plus a few nautiloids.

*Age:* Early Late Llandoverian (Rong and Yang, 1977; Ge and others, 1979).

*Typical Locality:* Dazhongba, Yichang County, western Hubei Province.

*Geographic Distribution:* Yichang area, western Hubei Province.

*Environment:* The articulated nature of the shells, together with the mudstone matrix is consistent with a relatively quiet-water environment, despite the presence of fairly abundant *Harpidium* and *Stricklandia.* There is a significant difference between a high-dominance, rough-water pentameroid community, and a high-diversity, quiet-water community in which pentameroids are only one of the taxa present. The composition of the fauna indicates a Benthic Assemblage 4 environment, i.e., absence of deeper water or quieter water taxa such as *Brevilamnulella* and *Cyrtia* in abundance. The presence of a stricklandid and *Harpidium,* and the absence of *Dicoelosia* and *Skenidioides,* indicate a position near the Benthic Assemblage 3-4 boundary.

*Community Group:* Pentamerinae, Stricklandiidae, and *Striispirifer* Community Group mixture.

### Howellella tingi Community (Fig. 11; Table 9)

*Name:* This community is named here. It is similar to Boucot's (1975, p. 261) "Tentaculitid Community I" in its abundance of *Howellella,* but the absence of abundant tentaculitids and *Leperditia* separates them.

*Composition:* Table 9 indicates a low- to medium-diversity community dominated by *Howellella tingi.* The two samples in which significant numbers of *Protathyrisina, Protathyris,* and *Eoschizophoria* are present may indicate boundary mixing between the *Howellella tingi* Community and the *Protathyrisina, Protathyris,* and *Eoschizophoria hesta–Protathyris xungmiaoensis* Communities, respectively.

*Age:* Late Ludlovian and Early Pridolian (Wang and others, 1980; Rong and Yang, 1980).

*Typical Locality:* Qujing, eastern Yunnan Province.

*Geographic Distribution:* Eastern Yunnan, only in the Miaokao Formation.

*Environment:* Some of the samples occur largely as articulated individuals in calcareous mudstones, whereas others occur in a largely disarticulated condition. The articulated condition of some of the samples is probably the normal situation, which leads us to conclude that the community represents a relatively quiet-water environment that would help to explain the relatively low diversity. The association is a Benthic Assemblage 2 community. Similar communities occur in some of the platy limestone of the Lochkovian in the Prague region where entire bedding planes are covered with *Howellella.*

*Community Group:* Striispirifer.

### Howellella–Reticulariopsis Community (Fig. 15)

*Name:* This unit is named here.

*Age:* Early Emsian or Zlichovian (see Wang Yu and others,

TABLE 9. HOWELLELLA TINGI COMMUNITY

| Name | Articulated Shell | Pedicle Valve | Brachial Valve | Either Valve | Total No. Specimens | % |
|---|---|---|---|---|---|---|
| **Locality ADH 55, Miaokao Formation** | | | | | | |
| Howellella tingi | 23 | 253 | 118 | 5 | 399 | 52 |
| Protathyrisina uniplicata | 8 | 2 | | | 10 | 1 |
| P. minor | 66 | 5 | | | 71 | 9 |
| P. plicata | 19 | | | | 19 | 3 |
| P. quadriplicata | 3 | | | | 3 | |
| P. sp. | 32 | 4 | 14 | 17 | 67 | 9 |
| Protathyrisina total | 128 | 11 | 14 | 17 | 170 | 22 |
| Eoschizophoria hesta | 13 | 82 | 21 | 33 | 149 | 20 |
| Aesopomum delicatum | | 3 | 2 | 2 | 7 | 1 |
| Protathyris planosulcata | 5 | | | | 5 | 1 |
| P. xungmiaoensis | 16 | 4 | | 3 | 23 | 3 |
| Spirinella asiatica | 3 | 5 | | | 8 | 1 |
| Lingula orientalis | 1 | | 1 | | 2 | |
| Total | | | | | 763 | |
| **Locality ADH 61, Miaokao Formation** | | | | | | |
| Howellella tingi | 841 | 171 | 36 | | 1,048 | 57 |
| Protathyrisina uniplicata | 282 | 4 | | | 286 | 15 |
| P. minor | 148 | 3 | 1 | | 152 | 8 |
| P. plicata | 8 | | | | 8 | |
| P. spp. | | | | 2 | 2 | |
| Protathyris xungmiaoensis | 264 | 2 | | | 266 | 14 |
| P. planosulcata | 86 | 2 | | | 88 | 5 |
| Total | | | | | 1,850 | |
| **Locality ADH 62, Miaokao Formation** | | | | | | |
| Howellella tingi | 269 | 164 | 66 | | 499 | 100 |
| Lingula orientalis | 1 | | | | 1 | |
| Total | | | | | 500 | |
| **Locality ADH 63, Miaokao Formation** | | | | | | |
| Howellella tingi | 17 | 81 | 47 | | 145 | 100 |
| Lingula sp. | | 1 | | | 1 | |
| Total | | | | | 146 | |
| **Locality ADH 66, Miaokao Formation** | | | | | | |
| Howellella tingi | | 20 | 30 | | 50 | 100 |
| **Locality ADG 67, Miaokao Formation** | | | | | | |
| Howellella tingi | 21 | 102 | 39 | 30 | 192 | 100 |

1979; Wang Cheng-yuan and others, 1979); this unit is dated by the presence of *Polygnathus perbonus*) (see also Yu and Yin, 1978, p. 24, for conodont data).

***Composition:*** The brachiopods in the Ertang Formation occur as scattered individuals in calcareous mudstone. The most common shells are *Howellella* and *Reticulariopsis,* but specimens of *Orientospirifer*(?), *Parathyrisina, Athyrisina,* and a chonetid also occur. The available samples are too small to provide meaningful statistics. However, about 300 specimens from a number of small collections demonstrate that this is a low-diversity community.

***Typical Locality:*** Ertang, Wuxuan County, central Guangxi.

***Geographical Distribution:*** Same as above.

***Environment:*** This community suggests a nearer shore environment than the contemporary *Vagrania-Leptathyris* Community. The lithology and articulated condition of many of the shells indicate fairly quiet water. The scattered condition of the fossils, taken together with the low-population density, suggests a restricted environment largely unfavorable to benthic shells. The beds are not thinly laminated, or even thinly bedded, so that a low-oxygen, quiet-water environment is considered unlikely. A Benthic Assemblage 2 or 3 environment is most likely in view of the paleogeographic position of the unit. Both lithofacies and biofacies evidence suggests that this Ertang Formation community represents a nearer shoreline environment than does the *Huananochonetes-Xenostrophia* Community found in the Liujing region to the southwest of Ertang itself. In the Ertang region, the Yukiang Formation (from the upper part of which in the Liujing area is found the *Huananochonetes-Xenostrophia* Community) is represented not only by a relatively nearshore marine and estuarine facies bearing some brachiopods but also by sandstone with

plants and vertebrate fossils (Yu and Yin, 1978). The Ertang Formation *Howellella-Reticulariopsis* Community represents a somewhat nearer shore environment than does the Yukiang Formation of the Liujing Area.
*Community Group: Striispirifer.*

### *Huananochonetes–Xenostrophia* Community (Fig. 15; Table 10)

*Name:* This unit is named here.
*Age:* The Liujing Member of the Yukiang Formation is of Early Emsian age (Ruan and others, 1979; Wang Cheng-yuan and others, 1979).
*Composition:* Table 10 provides data from four samples of this community. Dominant taxa are largely the same from collection to collection, but changes in abundance rank do take place: contrast the composition and relative abundance of the taxa in this community with those occurring in the underlying *Acrospirifer-Atrypa* Community of the Shizhou Member of the Yukiang Formation. Not all taxa are shared between the two communities and among the taxa that are shared, are important differences in relative abundance.
*Typical Locality:* Liujing, Hengxian County, southern Guangxi.
*Geographic Distribution:* South China, including Guangxi, northern Sichuan (Chen, 1979; Wan and others, 1978), and possibly northern Vietnam if the material published by Mansuy (1908) and Patte (1926) belongs here.
*Environment:* The commonly articulated nature of the brachiopods, taken together with the muddy matrix, suggests a relatively quiet-water environment. The high diversity is consistent with a Benthic Assemblage 3 assignment, as is the absence of Benthic Assemblage 4 and deeper water shells. The large size of many of the taxa is consistent with a Benthic Assemblage 3 rather than a 4-5 assignment.
*Community Group: Striispirifer.*
*Discussion:* Mr. Zhang Ning collected, sorted, and identified two bulk samples from ADH 17, one from ADH 18, and one from ADH 19. The second bulk sample from ADH 17 is three times as large as the first. The bulk samples consisted of "bagged" surface material scooped up to a depth of about 15 cm at the locality where the handpicked shells listed in the table were obtained. It should be noted that the percentage of smaller species is much higher in the bulk samples than in the handpicked ones, as would be expected from the earlier, similar sampling studies of Sparks (*in* Ager, 1963). For ADH 17, both bulk samples yielded similar results for the abundant taxa, but the larger sample produced more taxa, as would also be expected. For ADH 17, there are major differences in percentage between the large shells in the handpicked and bulk samples. It is clear that the collectors favored large, conspicuous shells lying on the surface rather than the smaller shells. However, only by handpicking would the rarer large species be obtained unless a prohibitively large bulk sample were processed. The large bulk sample from ADH 17 weighed about 2 kg.

### *Leiorhynchus* Community

*Name:* We do not formally propose this term here, but the concept of a high-dominance, low-diversity leiorhynchid community has long been cited in the literature (Savarese and others, 1986, is a typical example).
*Age:* Chen Yuan-ren (*in* Wan and others, 1978; written communication, 1980) assigns beds of the Puqiaozi Member of the Shawozi Formation, northern Sichuan, containing this community to the Frasnian (Xian and Jiang, 1978, report the same fauna in Puding and Anshan, Guizhou).
*Composition:* Abundant articulated specimens of *Leiorhynchus.*
*Typical Locality:* None designated.
*Geographic Distribution:* Worldwide, outside of the Malvinokaffric Realm, in the Middle and Upper Devonian.
*Environment:* The articulated shells occurring in abundance in a dark-colored limestone represent a Benthic Assemblage 5, possibly low-oxygen environment, as is characteristic of the community at many other places in the world.
*Community Group:* None designated.

### *Lingula* Community (Fig. 10)

*Name:* The term *Lingula* Community is used here in the sense of a very low-diversity community dominated by *Lingula,* and occurring in the Benthic Assemblage 1 position.
*Composition:* Scattered linguloids only.
*Age:* The age of the upper beds of the Fentou Formation, from which this unit has been obtained near Nanjing, is difficult to define. The highest graptolites in the underlying sequence indicate an early Late Llandoverian *M. sedgwickii* Zone age. Above the graptolite-bearing beds are lower beds of the Fentou Formation that contain a shelly fauna and vertebrates that by themselves provide only a generalized Llandoverian-Ludlovian age. But, due to the gradational relations with the underlying graptolitic beds, it is reasonable to conclude that these beds of the Fentou Formation, which are only some tens of meters thick, are of Late Llandoverian–Early Wenlockian age. The thin, overlying *Lingula* Community-bearing part of the Fentou Formation is also most reasonably assigned a Late Llandoverian–Early Wenlockian age. Overlying the Fentou Formation, with apparent conformity, are the unfossiliferous, cross-bedded, red beds of the Silurian Maoshan Formation (itself unconformably overlain by the Upper Devonian). The Maoshan lacks marine trace fossils such as *Scolithus* or *Arthrophycus* of Benthic Assemblage 1, and a nonmarine assignment is reasonable.
*Typical Locality:* None designated.
*Geographic Distribution:* Nanjing region.
*Environment:* The *Lingula* Community of the Fentou Formation, in the context of an upward-shallowing sequence, is entirely consistent with a Benthic Assemblage 1 assignment.
*Community Group:* A Linguloid Community Group should be erected for the varied, linguloid dominated Benthic Assemblage 1 communities such as this one, but only after these shells have received the taxonomic treatment they deserve.

TABLE 10. HUANANOCHONETES-XENOSTROPHIA COMMUNITY

| Name | Articulated Shell | Pedicle Valve | Brachial Valve | Either Valve | Total No. Specimens | % |
|---|---|---|---|---|---|---|
| **Locality ADH 16** | | | | | | |
| Huananochonetes | 32 | | | | 32 | 8 |
| Xenostrophia | 115 | 1 | 4 | | 120 | 31 |
| Levenea | 32 | 19 | 75 | | 126 | 32 |
| Acrospirifer | 15 | 6 | | | 21 | 5 |
| Parathyrisina | 10 | | | | 10 | 3 |
| Cyrtina | 1 | | | | 1 | - |
| Uncinulus | 2 | | | | 2 | 1 |
| Eosophragmophora | 5 | | 2 | | 7 | 2 |
| Dicoelostrophia | 34 | | | | 34 | 9 |
| Leptaenopyxis | | | | 10 | 10 | 3 |
| "Cymostrophia" | | | | 1 | 1 | - |
| Stropheodontid | | | 1 | | 1 | - |
| "Chonetes" kwangsiensis | 12 | | | | 12 | 3 |
| Schizophoria | 4 | 1 | | | 5 | 1 |
| Athyris | 2 | | | | 2 | 1 |
| Elymospirifer | 1 | | 1 | | 2 | 1 |
| Glyptospirifer | 2 | | 1 | | 3 | 1 |
| "Megastrophia" | 2 | | | | 2 | 1 |
| Productella or Spinulicosta | | | | 1 | 1 | - |
| Total | | | | | 392 | |
| **Locality ADH 17** | | | | | | |
| Xenostrophia | 191 | | | | 191 | 43.0 |
| Huananochonetes | 67 | 11 | 4 | | 82 | 18.5 |
| Eosophragmophora | 48 | 12 | 8 | | 68 | 15.0 |
| Uncinulus | 33 | | | | 33 | 8.0 |
| Rostrospirifer | 20 | 4 | 2 | | 26 | 5.8 |
| Acrospirifer | 5 | 2 | 2 | | 9 | 2.0 |
| Parathyrisina | 6 | | | | 6 | 1.4 |
| Elymospirifer | 1 | | | | 1 | 0.2 |
| Levenea | 2 | | | | 2 | 0.5 |
| Dicoelostrophia | 10 | 1 | | | 11 | 2.5 |
| Athyris grandis | 1 | | | | 1 | 0.2 |
| A. sp. | 1 | | | | 1 | 0.2 |
| "Cymostophia" | 1 | | | | 1 | 0.2 |
| "Megastrophia" | 1 | | | | 1 | 0.2 |
| "Chonetes" kwangsiensis | 8 | 1 | | | 9 | 2.0 |
| Total | | | | | 442 | |
| **Locality ADH 18** | | | | | | |
| Xenostrophia | 152 | | 2 | | 154 | 14.7 |
| Huananochonetes | 258 | | 1 | 18 | 277 | 26.0 |
| Uncinulus | 177 | | 3 | | 180 | 17.0 |
| Rostrospirifer | 141 | | 10 | | 151 | 14.0 |
| "Chonetes" kwangsiensis | 157 | | | | 157 | 14.9 |
| Dicoelostrophia | 40 | 1 | | | 41 | 3.9 |
| Parathyrisina | 7 | | | | 7 | 0.6 |
| "Megastrophia" | 7 | | | | 7 | 0.6 |
| Glyptospirifer | 6 | | | | 6 | 0.6 |
| Leptaenopyxis | 2 | | 1 | | 3 | 0.3 |
| Athyris | 3 | | | | 3 | 0.3 |
| Elymospirifer | | 2 | | | 2 | 0.2 |
| Levenea | 12 | 15 | 14 | | 41 | 4.0 |
| "Cymostrophia" | 2 | | | | 2 | 0.2 |
| Eosophragmophora | 18 | 1 | 1 | | 20 | 2.0 |
| Total | | | | | 1,051 | |

TABLE 10. HUANANOCHONETES-XENOSTROPHIA COMMUNITY (continued)

| Name | Articulated Shell | Pedicle Valve | Brachial Valve | Either Valve | Total No. Specimens | % |
|---|---|---|---|---|---|---|
| **Locality ADH 19** | | | | | | |
| Huananochonetes | 352 | | 2 | | 354 | 47.7 |
| Xenostrophia | 93 | | 19 | | 112 | 15.0 |
| "Chonetes" kwangsiensis | 36 | | 2 | | 38 | 5.0 |
| Dicoelostrophia | 35 | 5 | 3 | | 43 | 5.8 |
| Acrospirifer | 34 | 11 | | | 45 | 6.0 |
| Rostrospirifer | 26 | 3 | | | 29 | 3.9 |
| Levenea | 9 | 41 | 39 | | 89 | 10.6 |
| Eosophragmophora | 17 | 2 | 1 | | 20 | 2.6 |
| Uncinulus mesodeflectus | 10 | | | | 10 | 1.3 |
| U. fasciger | 4 | | | | 4 | 0.5 |
| Athyris grandis | 1 | 2 | | | 3 | 0.4 |
| "Megastrophia" | 1 | | | | 1 | 0.1 |
| Cyrtina | 3 | | | | 3 | 0.4 |
| Total | | | | | 751 | |
| **ADH 17-Bulk Sample 1 (1 bag)** | | | | | | |
| Huananochonetes | 88 | | 1 | | 89 | 35.0 |
| Xenostrophia | 13 | 2 | 3 | | 18 | 7.0 |
| Eosophragmophora | 34 | 61 | 34 | | 121 | 50.0 |
| Plicate spiriferid fragments | | 5 | 1 | | 12 | 7.0 |
| Total | | | | | 240 | |
| **ADH 17-Bulk Sample 3x (3 bags)** | | | | | | |
| Huananochonetes | 232 | | 2 | | 234 | 38.0 |
| Xenostrophia | 42 | | | | 42 | 6.8 |
| Eosophragmophora | 137 | 114 | 57 | | 308 | 50.5 |
| Plicate spiriferid fragments | 1 | 10 | 5 | 4 | 20 | 3.0 |
| Dicoelostrophia | 3 | | | | 3 | 0.5 |
| Uncinulus | 1 | | | | 1 | 0.2 |
| Leptaenopyxis | 1 | | | | 1 | 0.2 |
| Total | | | | | 609 | |
| **ADH 18-Bulk Sample (3 bags)** | | | | | | |
| Huananochonetes | 13 | | | | 13 | 62.0 |
| Plicate spiriferid fragments | | 3 | 2 | | 5 | 24.0 |
| Eosophragmophora | 1 | | | | 1 | 5.0 |
| Uncinulus | 1 | | | | 1 | 5.0 |
| Dicoelostrophia | 1 | | | | 1 | 5.0 |
| Total | | | | | 21 | |
| **ADH 19-Bulk Sample (2 bags)** | | | | | | |
| Huananochonetes | 75 | | | | 75 | 48.0 |
| Eosphragmophora | 15 | 32 | 25 | | 72 | 46.0 |
| Xenostrophia | 4 | | | | 4 | 2.5 |
| Plicate spiriferid fragments | 2 | | | | 2 | 1.3 |
| Atrypa | 1 | | | | 1 | 0.5 |
| Rhynchonellid | 1 | | | | 1 | 0.5 |
| Plectodonta | 1 | | | | 1 | 0.5 |
| Total | | | | | 156 | |

TABLE 10. HUANANOCHONETES-XENOSTROPHIA COMMUNITY (continued)

| Name | Articulated Shell | Pedicle Valve | Brachial Valve | Either Valve | Total No. Specimens | % |
|---|---|---|---|---|---|---|
| *Upper Part of Liujing Member at Side of Road near Commune* | | | | | | |
| Huananochonetes | 106 | | 1 | | 107 | 46 |
| Eosophragmophora | 69 | 20 | 18 | | 107 | 46 |
| Acrospirifer | 2 | 2 | 1 | | 5 | 2 |
| Rostrospirifer | 3 | | | | 3 | 1 |
| Dicoelostrophia | 4 | | | | 4 | 2 |
| "Chonetes" kwangsiensis | 4 | | | | 4 | 2 |
| Uncinulus | 1 | | | | 1 | |
| Total | | | | | 231 | |
| *Top of Liujing Member behind Warehouse* | | | | | | |
| Acrospirifer | 26 | 7 | | | 33 | 13 |
| "Chonetes" kwangsiensis | 28 | | 8 | | 36 | 15 |
| Xenostrophia | 24 | | | | 24 | 9 |
| Eosophragmophora | 13 | 5 | 1 | | 19 | 7 |
| Rostrospirifer | 13 | | | | 13 | 5 |
| Dicoelostrophia | 10 | | 1 | | 11 | 4 |
| Huananochonetes | 10 | | | | 10 | 4 |
| "Megastrophia" | 2 | | | | 2 | 1 |
| Parathyrisina | 10 | | | | 10 | 4 |
| Schizophoria | 2 | | | | 2 | 1 |
| Glyptospirifer | 5 | | | | 5 | 2 |
| Howellella sp. | 2 | | | | 2 | 1 |
| Latonotoechia | 5 | | | | 5 | 2 |
| Uncinulus | 83 | | | | 83 | 32 |
| Athyris grandis | 1 | | | | 1 | – |
| Total | | | | | 256 | |

*Nowakia* **Community (Pelagic Community) (Figs. 16 and 17)**

*Name:* We do not formally assign a name to this unit for either the Emsian or the Eifelian of the Nandan Formation in South China. It is part of the same basic unit discussed by Boucot (1975) as the Pelagic Community. Goniatites are commonly present as well as different species of *Nowakia.*

*Age:* Emsian and Eifelian (see Wang Cheng-yuan and others, 1979).

*Composition:* Abundant *Nowakia,* covering bedding planes.

*Geographic Distribution:* Guangxi Province, and adjacent parts of North Vietnam (Mansuy, 1908; Patte, 1926). Also present in the black shale facies of the Middle Devonian in eastern North America (the *Styliolina* Facies), and in the Wissenbacher Schiefer of the Rhineland (Langenstrassen, 1972).

*Typical Locality:* None designated.

*Environment:* The occurrence of this community in thin-bedded to laminated, nonbioturbated, dark-colored mudstone and shale, commonly with no benthic shells present, although goniatites are commonly found, is good evidence for a fairly anoxic bottom onto which pelagic shells such as *Nowakia* and goniatites fell after death. The common current alignment of the *Nowakia* indicates, a certain level of current activity; this has been observed elsewhere in the world. These beds commonly indicate sites of low sedimentation rate.

In the Liujing area, thin-bedded limestones containing the *Nowakia* Community interfinger with Benthic Assemblage 3 strata containing *Stringocephalus.* Near this boundary, scattered, articulated *Stringocephalus* occur in the platy, dark-colored *Nowakia*-bearing limestone. Consequently, the *Nowakia* Community does occur near the Benthic Assemblage 3 boundary if not actually landward of it.

In the Daliancun Member of the Yukiang Formation at Daliancun, near Nanning, Guangxi, Ruan Yi-ping (oral communication, 1980) reported *Nowakia* in argillaceous limestone associated with *Eosophragmophora,* a Benthic Assemblage 3 brachiopod. This occurrence further documents the presence of *Nowakia* into the Benthic Assemblage 3 environment, although this is not the normal kind of occurrence.

*Community Group:* Not enough is known about the community ecology of the tentaculitids to permit the erection of community groups at this time.

*Nucleospira–Nalivkinia* **Community (Fig. 10)**

*Name:* This community is named here.

*Composition:* Scattered specimens of both *Nucleospira* and *Nalivkinia, Striispirifer* of the *Hedeina* type, *Hunanodendrum*-type dendroids, and a few other taxa, are found occasionally. At a few localities, the dominance of *Nucleospira* is such that almost the entire community is composed of it.

*Age:* About mid-Upper Llandoverian following Ge and others (1979).

*Geographic Distribution:* Northern and northeastern Guizhou,

TABLE 11. ORIENTOSPIRIFER-SINOCHONETES COMMUNITY

| Name | Articulated Shell | Pedicle Valve | Brachial Valve | Either Valve | Total No. Specimens | % |
|------|------|------|------|------|------|------|
| **Locality ADH 6** | | | | | | |
| "*Orientospirifer*" *wangi* | 1 | 1 | | | 2 | 1 |
| *Orientospirifer nakaolingensis* | 7 | 53 | 51 | 38 | 149 | 78 |
| *Protathyris praecursor* | 13 | | | | 13 | 7 |
| *Howellella* sp. | | 5 | 2 | 4 | 11 | 6 |
| *Sinochonetes minutisulcatus* | | 11 | 1 | 3 | 15 | 8 |
| *Kwangsirhynchus liujingensis* | 1 | | | | 1 | 1 |
| Total | | | | | 191 | |
| **Locality ADH 7** | | | | | | |
| "*Orientospirifer*" *wangi* | 6 | 8 | 6 | | 20 | 10 |
| *Orientospirifer nakaolingensis* | 17 | 7 | | | 24 | 12 |
| *Orientospirifer* n. sp. | 43 | 37 | 11 | | 91 | 44 |
| "*Orientospirifer*" sp. | 1 | | | | 1 | 1 |
| *Aseptalium guangxiense* | 8 | 2 | 4 | | 14 | 7 |
| *Sinochonetes minutisulcatus* | 35 | 10 | 12 | | 57 | 27 |
| *Protathyris praecursor* | | 2 | | | 2 | 1 |
| Total | | | | | 209 | |

southeastern Sichuan, southwestern Hubei, and northwestern Hunan. Well over 20 localities.

*Environment:* This community commonly occurs in argillaceous siltstone, argillaceous sandstone, and sandy shale. The shells are commonly disarticulated and scattered. The community has been found in the Majiaochong, Rongxi, Hanchiatien, and Lower Xiushan Formations. Marine redbeds with ripple marks, cross-bedding and a single taxon bivalve assemblage are sometimes interbedded with this community in the Rongxi Formation. The stratigraphic position of this community in the upper part of the generally upward-shallowing Silurian of southern China, the very low-diversity fauna with scattered shells, the presence of a dendroid, and dominant *Nucleospira* are all consistent with a restricted, very shallow, Benthic Assemblage 2 position (Rong and others, 1984).

*Community Group: Striispirifer.*

*Discussion:* This unit is distinct from Boucot's (1975) *Nalivkinia* Community, which is a low-diversity, high-dominance, moderately high-population density, Benthic Assemblage 2 unit in which most of the shells are articulated. This unit also differs from a few calcareous mudstone collections from the Leijiatun Formation (Zone of *M. sedgwickii*) in Shiqian County, northeastern Guizhou, at Leijiatun, which consist largely of articulated specimens of *Nalivkinia* associated with abundant rugose corals; this suggests a Benthic Assemblage 3 assignment.

*Orientospirifer–Sinochonetes* Community (Fig. 14; Table 11)

*Name:* This community is named here.

The term "Chonetid Community" was used previously by Boucot (1975, p. 247) to identify Rhenish-Bohemian Region Lower Devonian high-dominance, low-diversity Benthic Assemblage 2 units dominated either by *Chonetes* or by *Plebejoch-*

*onetes.* Feldman (1980) used the term to refer to an Onondaga Limestone community from New York in which the dominant chonetid was not generically identified. The Rhenish-Bohemian Region communities should be termed the *Chonetes* and *Plebejochonetes* Communities, respectively. We recommend that the term "Chonetid Community" be reserved for situations in which the dominant chonetid species cannot be generically identified, i.e., when dealing with relatively indeterminable specimens.

*Age:* Middle-Late Siegenian based on the presence of *Eognathus sulcatus* and *Polygnathus* sp. (cf. *dehiscens*) (Ruan and others, 1979; Wang Cheng-yuan and others, 1979) in the Nahkaoling Formation.

*Composition:* This community (Table 11) is dominated by various forms of *Orientospirifer* together with abundant *Sinochonetes,* and fairly abundant specimens of *Aseptalium.* There are also uncommon specimens of *Protathyris* and rarer specimens of *Kwangsirhynchus, Atrypa* "*reticularis,*" *Howellella,* an orthotetacid, and a stropheodontid. The bivalves *Paracyclas* and *Eoschizodus* occur in low frequency. Tentaculitids, chitinozoans, acritarchs, and plant spores are also present (microfossils cited by Hou, 1978, and Gao, 1978).

*Typical Locality:* Liujing, Hengxian County, South Guangxi.

*Geographic Distribution:* In the Liujing region of Guangxi, and possibly in the Heigou Member, Bailiuping Formation, from Bailiuping, Ganxi, Beichuan County, N. Sichuan (Chen, 1979).

*Environment:* Most specimens occur as articulated shells in a marly matrix, or as disarticulated specimens in a muddy matrix containing a large percentage of sand-sized materials. This low- to medium-diversity community, rich in tentaculitids, represents a Benthic Assemblage 2, relatively quiet-water nearshore environment. This interpretation is consistent with the following factors: the presence of Benthic Assemblage 1 Bivalve communities in the underlying Lianhuashan Formation, a varied vertebrate fauna as well as plant spores, and the presence of many Benthic Assemblage 3 communities in the overlying Yukiang Formation. This interpretation is consistent with the presence of interlayered *Protathyris* Community that also belongs to Benthic Assemblage 2.

Hou (1978) listed the following chitinozoans from the Nahkaoling Formation: abundant *Angochitina (A. devonica, A. liujingensis),* plus *Ancyrochitina ancyrea, A. tumida, A. spinosa* (*Angochitina* makes up about 60-75 percent of the assemblage).
***Community Group:*** *Striispirifer.*

### *Paraconchidium–Virgianella* Community (Fig. 10)

***Name:*** The *Paraconchidium-Virgianella* Community was defined by Rong and Yang (1981).
***Composition:*** A single taxon community, with the exception that a few specimens of *Virgianella* are present from place to place.
***Age:*** The beds yielding this community; are probably of very early Late Llandoverian age (Rong and Yang, 1981).
***Typical Locality:*** Juntianba, Shiqian, northeastern Guizhou.
***Geographic Distribution:*** Northeastern Guizhou; about 11 localities are known (Ge and others, 1979); see *Spirigerina* Community for more details.
***Environment:*** The *Paraconchidium–Virgianella* Community occurs in a 2.5-m-thick bed of grayish-black limestone at the typical locality. Most of the specimens are disarticulated. This is a typical rough-water, Benthic Assemblage 3, single-taxon, pentameroid community.
***Community Group:*** Virgianidae.
***Discussion:*** Rong and Yang (1981) defined the community as one in which *Paraconchidium* is dominant and *Virgianella* is rare. See the section entitled "Naming of Communities" for a discussion of some of the problems involved in selecting community names.

### *"Paracraniops"-Paromalomena* Community (Fig. 7; Table 12)

***Name:*** Rong (1979) used the term *Draborthis-Toxorthis* Assemblage for what we here term the *"Paracraniops"-Paramalomena* Community. He used the term Assemblage in a manner similar to using the term faunule rather than Community.
***Composition:*** Rong (1979) discussed the brachiopods of this community (see also Table 12); in addition, there are a number of other taxa present of which the most conspicuous are *Dalmanitina* and a hyolithid. The brachiopod genera include those present in the allied *Aphanomena-Hirnantia* and *Aphanomena-Hirnantia-Dalmanitina* communities, plus a large number of additional brachiopod genera and a hyolithid. Relative abundances of the genera *Aphanomena, Hirnantia, Paromalomena,* and *Dalmanitina* are lower in the *"Paracraniops"-Paromalomena* Community than in the others. This distinct difference in relative abundance is consistent with the normal situation in which dominance of any taxon within a community generally decreases as overall diversity increases.

***Age:*** This unit occurs in beds of Ashgillian age (Rong, 1979) lying above the Ashgillian-age graptolite Zone of *Diplograptus bohemicus,* present in the underlying Wufeng Formation.
***Typical Locality:*** Rong proposes that Locality WM 188 (Tangya, near Yichang, Hubei) be the typical locality.
***Geographic Distribution:*** Rong (1979) recognized this community at about eight localities in Sichuan and Hubei Provinces.
***Environment:*** The shells of this community occur as fairly closely packed individuals lying on single bedding planes rather than in shell beds. Most of the shells are disarticulated and breakage is minimal. Most of the convex shells lie with the convex side up. The associated rock is a very calcareous mudstone.

The paleogeographic and stratigraphic positions lead Rong to conclude that this community occurs seaward of both the *Aphanomena-Hirnantia* and *Aphanomena-Hirnantia-Dalmanitina* Communities. The higher diversity of the *"Paracraniops"-Paromalomena* Community is consistent with the presence of a normal, not too turbulent, marine environment. However, the overall character of the community is consistent with a Benthic Assemblage 3 rather than 4 or 5 assignment because of the absence of such deeper water taxa as *Dicoelosia, Skenidioides, Brevilamnulella, Plectodonta, Leangella,* and *Foliomena.*

Biogeographically, this community is similar to both the *Aphanomena-Hirnantia* and *Aphanomena-Hirnantia-Dalmanitina* Communities. The presence of a fairly common, smooth hyolithid is consistent with the presence of relatively cool water, as in the Malvinokaffric Realm Devonian (Boucot, 1975) where hyolithids and conularids are abundant.
***Community Group:*** We cannot now assign most Ordovician communities to community groups, owing to lack of study.

### *Paromalomena–Aegiromena* Community (Fig. 7)

***Name:*** This community corresponds to Rong's (1979) "Assemblage" of the same name, although he used it in the sense of "faunule."
***Composition:*** This community consists of relatively scattered specimens of *Paromalomena, Aegiromena,* and *Fardenia* (or *Coolinia*) together with a number of graptolites, a few ostracodes, bivalves and gastropods, a eurypterid, and *Dalmanitina.* The available brachiopod sample is small.
***Age:*** Rong (1979) recorded this community only from beds of Late Ashgillian age.
***Typical Locality:*** Xinkailing is designated as the typical locality.
***Geographic Distribution:*** Rong (1979) reported this community from two localities; one at Xinkailing, near Wuning, Jiangxi Province, and the other at Beigongli, Yunling, Jingxian County, Anhui Province.
***Environment:*** The community occurs in a thin 15–cm unit, the Xinkailing Bed, with underlying Wufeng Formation graptolitic shale and overlying graptolitic, gray-black mudstone of the Lishuwuo Formation. The shells in the Xinkailing Bed are disarticulated and occur in a black mudstone.

It is difficult to decide how best to assign this community.

TABLE 12. "PARACRANIOPS"-PAROMALOMENA COMMUNITY

| Name | Articulated Shell | Pedicle Valve | Brachial Valve | Either Valve | Total No. Specimens | % |
|---|---|---|---|---|---|---|
| **Locality WM 188** | | | | | | |
| "Paracraniops" | | | | 107 | 107 | 10 |
| Philhedra | | | | 2 | 2 | |
| Toxorthis | 2 | 4 | | 2 | 8 | 1 |
| Dalmanella | | 14 | 22 | 6 | 42 | 4 |
| Kinnella | | 27 | 36 | 23 | 86 | 8 |
| Draborthis | | | 3 | 1 | 4 | |
| Horderleyella | | 13 | 8 | 7 | 28 | 3 |
| Hirnantia | | 49 | 66 | 130 | 245 | 22 |
| Bancroftina? | | 1 | 3 | 1 | 5 | |
| Cliftonia | | 13 | 16 | 36 | 65 | 6 |
| Triplesia | | 11 | 8 | 42 | 61 | 5 |
| Aegiromena | | 27 | 8 | 4 | 39 | 4 |
| Paromalomena | | 58 | 44 | 92 | 194 | 17 |
| Aphanomena | 1 | 39 | 29 | 155 | 224 | 20 |
| Leptaenopoma | | 3 | | 2 | 5 | |
| Dorytreta | 2 | 1 | 2 | 4 | 9 | 1 |
| Plectothyrella | | 3 | 1 | | 4 | |
| Hindella | 2 | | | | 2 | |
| Total | | | | | 1,130 | |

| | Pygidium | Cephelon | | | | |
|---|---|---|---|---|---|---|
| Dalmanitina | 44 | 35 | | | 79 | |
| Lophospira? | | | | 1 | 1 | |
| Bellerophontid | | | | 4 | 4 | |
| Crinoids | Rare | | | | | |
| Hyolithid (small) | | | | 53 | 53 | |
| Hyolithid (large) | | | | 5 | 5 | |
| Total | | | | | 1,272 | |

The small size of the two samples, and the similarity of the brachiopods and trilobite to the other Ashgillian shelly communities precludes certainty. However, the more scattered nature of the fossils in their matrix, as well as the thin nature of the Xinkailing Bed, are inconsistent with an environment similar to those present in the other Benthic Assemblage 3 South China shelly Ashgillian communities. Using paleogeographic evidence, Rong concluded that this community lived much closer to the shoreline than the *Aphanomena-Hirnantia* and *"Paracraniops"-Paromalomena* Communities, but farther offshore than the *Aphanomena-Hirnantia-Dalmanitina* Community. Some sort of specialized Benthic Assemblage 3 environment seems called for, but its nature is unclear.

***Community Group:*** We are not now in a position to assign most Ordovician communities to community groups.

### *Plectodonta–Reticulariopsis* Community (Fig. 17)

***Name:*** This community is named here.

***Age:*** Eifelian (see Wang Yu and others, 1979; Wang Cheng-yuan and others, 1979). Abundant specimens of *Nowakia sulcata,* and rare goniatites, provide the age information (Xu Han-kui, oral communication, 1980) for this Nandan Formation community.

***Composition:*** Xu (1977, 1979) described the fauna (see Xu, 1979, Table 1, for a faunal list). He reported (oral communication, 1980) that *Plectodonta* is the most abundant brachiopod, followed by *Reticulariopsis* and *Costanoplia*. The notanopliids *Luofuia, Paraplicanoplia,* and *Tangxiangia* are rare, and the plectodontid *Nabiaoia* is very rare. Bivalves, trilobites, and goniatites are rare, but *Nowakia sulcata* is abundant.

***Typical Locality:*** Tangxiang-Nabiao Section, Luofu, Nandan County, northern Guangxi.

***Geographic Distribution:*** Same as above. A similar community may occur in Lingshan County, southern Guangxi.

***Environment:*** Dr. Xu stated (oral communication, 1980) that the brachiopods occur as very scattered, chiefly disarticulated, small specimens. The beds consist of dark-colored, thin-bedded (about 0.5 cm thick, on the average; Xu Han-kui, oral communication, 1980) mudstone. The *Nowakia* and goniatites are pelagic elements, and the trilobites may be epibenthic as these beds have not yielded trilobite trace fossils. We interpret these strata as Benthic Assemblage 4-5 strata, deposited in a relatively quiet-water, lower oxygen environment, as indicated by the thin-bedded, nonbioturbated nature of the strata. This community has much in common with the *Maoristrophia* Community of eastern Australia (Boucot, 1975), and does not contain a high enough

TABLE 13. PROTATHYRIS COMMUNITY

| Name | Articulated Shell | Pedicle Valve | Brachial Valve | Either Valve | Total No. Specimens | % |
|---|---|---|---|---|---|---|
| **Locality ADH 60, Miaokao Formation** | | | | | | |
| Protathyris xungmiaoensis | 697 | 25 | 1 | | 723 | 71 |
| P. planosulcata | 211 | | | 3 | 214 | 21 |
| Howellella tingi | 56 | 15 | | | 71 | 7 |
| Protathyrisina uniplicata | 6 | | | | 6 | 1 |
| P. plicata | 2 | | | | 2 | |
| P. sp. | | 1 | | | 1 | |
| Lingula orientalis | 1 | | | | 1 | |
| Total | | | | | 1,018 | |

TABLE 14. PROTATHYRISINA-GYPIDULA COMMUNITY

| Name | Articulated Shell | Pedicle Valve | Brachial Valve | Either Valve | Total No. Specimens | % |
|---|---|---|---|---|---|---|
| **Locality H 13** | | | | | | |
| Protathyrisina | 210 | | | | 210 | 76 |
| Schizophoria | 1 | | | | 1 | |
| Gypidula | 18 | | | | 18 | 7 |
| Howellella | 15 | | | | 15 | 5 |
| Stegerhynchus | 17 | | | | 17 | 6 |
| Atrypoidea | 8 | | | | 8 | 3 |
| Linguopugnoides | 4 | | | | 4 | 2 |
| Linguloid | 1 | | | | 1 | |
| Eospirifer | 1 | | | | 1 | |
| Stropheodontid | 1 | | | | 1 | |
| Total | | | | | 276 | |
| | | | | | | |
| Gastropods | 1 | | | | | |
| Trilobites | Rare | | | | | |

percentage of notanopliids to be put into the Notanopliid Community category, i.e., 90 percent or more notanopliids.
***Community Group:*** *Dicoelosia-Skenidioides.*

***Protathyris*** **Community (Figs. 11 and 14; Table 13)**

***Name:*** Boucot (1975) defined the *Protathyris* Community based chiefly on Appalachian materials. The Yunnan Silurian examples differ in some respects, but in the present state of our knowledge, we prefer to use the term *Protathyris* Community rather than devise a new one. The Lower Devonian Nahkaoling Formation materials discussed here are ecologically very similar to those Boucot discussed (1975).
***Composition:*** Abundant, dominant *Protathyris* (Table 13) with variable, lesser numbers of *Protathyrisina, Eoschizophoria, Tadschikia, Howellella,* and *Aesopomum* in the Silurian material. One hundred percent articulated specimens of moderate size *Protathyris* in the Devonian material.
***Age:*** Wang and others (1980) reported *crispus* Zone, Late Ludlovian age conodonts from the lower part of the Miaokao Formation. Siegenian (see *Orientospirifer-Sinochonetes* Community) for the Devonian material.

***Typical Locality:*** None chosen by Boucot (1975).
***Geographic Distribution:*** Liujing, Hengxian County, southern Guangxi, in the Nahkaoling Formation. Eastern Yunnan in the Miaokao Formation.
***Environment:*** The articulated shells, the dominance of the name bearer, and the fine-grained limestone matrix are all consistent with lower than normal oxygen and low current activity.
***Community Group:*** *Striispirifer.*

***Protathyrisina–Gypidula*** **Community (Fig. 12; Table 14)**

***Name:*** This community is named here.
***Composition:*** This is a medium-diversity community with a fairly high dominance of *Protathyrisina* accompanied by lesser numbers of *Gypidula, Howellella, Schizophoria,* a flat stropheodontid, *Linguopugnoides, Stegerhynchus,* and a linguloid (Table 14).
***Age:*** The age is uncertain, but somewhere in the range of Late Ludlovian to Late Pridolian is reasonable (Rong Jia-yu, oral communication, 1980).
***Typical Locality:*** Daerhanmaomingan, Inner Mongolia.
***Geographic Distribution:*** Same as above.

TABLE 15. PROTATHYRISINA PLICATA, P. MINOR, AND P. UNIPLICATA COMMUNITY

| Name | Articulated Shell | Pedicle Valve | Brachial Valve | Either Valve | Total No. Specimens | % |
|---|---|---|---|---|---|---|
| **Locality ADH 50, Miaokao Formation** | | | | | | |
| Protathyrisina uniplicata | 16 | 9 | 2 | | 27 | |
| P. minor | 2 | | | | 2 | |
| P. plicata | 5 | 1 | | | 6 | |
| P. spp. | 2 | 4 | | 16 | 22 | |
| Protathyrisina totals | 25 | 15 | 2 | 16 | 57 | 100 |
| **Locality ADH 51, Miaokao Formation** | | | | | | |
| Protathyrisina uniplicata | 81 | 5 | | | 86 | 10 |
| P. minor | 314 | 9 | | | 323 | 38 |
| P. plicata | 299 | 7 | | | 306 | 36 |
| P. quadriplicata | 21 | | | | 21 | 3 |
| P. spp. | 89 | 8 | 9 | 2 | 108 | 13 |
| Protathyrisina totals | 804 | 29 | 9 | 2 | 844 | 100 |
| Howellella tingi | 7 | 2 | | | 9 | 1 |
| Eoschizophoria hesta | 1 | | | | 1 | 0 |
| Protathyris sp. | 21 | | | | 21 | 2 |
| Total | | | | | 875 | |
| **Locality ADH 52, Miaokao Formation** | | | | | | |
| Protathyrisina uniplicata | 69 | 3 | 1 | | 73 | |
| P. minor | 191 | 7 | | | 198 | |
| P. plicata | 166 | 7 | | | 173 | |
| P. quadriplicata | 22 | 1 | | | 23 | |
| P. spp. | 47 | 2 | 5 | 4 | 58 | |
| Protathyrisina totals | 495 | 20 | 6 | 4 | 525 | 78 |
| Protathyris xungmiaoensis | 93 | 4 | | | 97 | |
| P. planosulcata | 31 | 1 | | | 32 | |
| Protathyris totals | 124 | 5 | | | 129 | 19 |
| Howellella tingi | 6 | 1 | 1 | | 8 | 1 |
| Eoschizophoria hesta | 8 | | | | 8 | 1 |
| Spirinella asiatica | 1 | | | | 1 | |
| Total | | | | | 671 | |
| **Locality ADH 53, Miaokao Formation** | | | | | | |
| Protathyrisina uniplicata | 9 | | | | 9 | |
| P. minor | 241 | 1 | 2 | | 244 | |
| P. quadriplicata | 3 | | | | 3 | |
| P. plicata | 166 | 2 | 1 | | 169 | |
| P. spp. | 23 | 17 | 3 | 4 | 47 | |
| Protathyrisina totals | 442 | 20 | 6 | 4 | 472 | 92 |
| Protathyris sp. | 1 | | | | 1 | |
| Howellella tingi | 4 | 2 | | | 6 | 1 |
| Eoschizophoria hesta | 19 | 2 | | | 21 | 4 |
| Spirinella asiatica | 6 | 3 | 1 | | 10 | 2 |
| S. biplicata | 4 | 1 | | | 5 | 1 |
| Total | | | | | 515 | |
| **Locality ADH 54, Miaokao Formation** | | | | | | |
| Protathyrisina uniplicata | 3 | 3 | | | 6 | |
| P. minor | 97 | 5 | | 3 | 105 | |
| P. plicata | 424 | 195 | 25 | 90 | 734 | |
| P. quadriplicata | 19 | 2 | | 1 | 22 | |
| P. spp. | 33 | 303 | 41 | 171 | 548 | |
| Protathyrisina totals | 576 | 508 | 66 | 265 | 1,415 | 99 |
| Protathyris planosulcata | 3 | | | | 3 | |
| P. xungmiaoensis | 3 | | | | 3 | |
| Eoschizophoria hesta | 1 | 1 | | 1 | 3 | |
| Total | | | | | 1,424 | |

TABLE 15. PROTATHYRISINA PLICATA, P. MINOR, AND P. UNIPLICATA COMMUNITY (continued)

| Name | Articulated Shell | Pedicle Valve | Brachial Valve | Either Valve | Total No. Specimens | % |
|------|-----------------|---------------|----------------|--------------|--------------------|----|
| **Locality ADH 56, Miaokao Formation** | | | | | | |
| Protathyrisina minor | 485 | 5 | | | 490 | |
| P. plicata | 131 | 3 | | | 134 | |
| P. uniplicata | 47 | 3 | | | 50 | |
| P. sp. | 44 | 2 | | 11 | 57 | |
| Protathyrisina totals | 707 | 13 | | 11 | 731 | 69 |
| Lingula sp. | 1 | 1 | 1 | | 3 | |
| Eoschizophoria hesta | 70 | 36 | | 3 | 109 | 10 |
| Aesopomum delicatum | 3 | | | 1 | 4 | |
| Atrypoidea inflata | 14 | 2 | | | 16 | 2 |
| Protathyris planosulcata | 66 | | | | 66 | 6 |
| P. xungmiaoensis | 66 | 8 | | | 74 | 7 |
| Howellella tingi | 22 | 27 | 16 | | 65 | 6 |
| Total | | | | | 1,068 | |
| **Locality ADH 58, Miaokao Formation** | | | | | | |
| Protathyrisina uniplicata | 29 | | | | 29 | |
| P. minor | 456 | 2 | | | 458 | |
| P. plicata | 21 | | | | 21 | |
| P. quadriplicata | 5 | | | | 5 | |
| P. sp. | 6 | | | | 6 | |
| P. puta | 2 | | | | 2 | |
| Protathyrisina totals | | | | | 521 | 50 |
| Eoschizophoria hesta | 173 | 119 | 5 | 158 | 455 | 43 |
| Lingula orientalis | 21 | | | 4 | 25 | 2 |
| Protathyris xungmiaoensis | 5 | | | | 5 | 1 |
| P. planosulcata | 3 | | | | 3 | 0 |
| Spirinella sp. | 5 | | | | 5 | 1 |
| Orbiculoidea? sp. | 1 | | | | 1 | 0 |
| Aesopomum delicatum | 1 | 8 | 8 | 5 | 22 | 2 |
| Howellella tingi | 9 | 3 | | | 12 | 1 |
| Total | | | | | 1,049 | 100 |
| **Locality ADH 57, Miaokao Formation** | | | | | | |
| Protathyrisina uniplicata | 23 | | | | 23 | |
| P. minor | 539 | | | | 539 | |
| P. plicata | 193 | | | | 193 | |
| P. quadriplicata | 7 | | | | 7 | |
| P. sp. | 29 | | | 1 | 30 | |
| Protathyrisina totals | | | | | 792 | 52 |
| Protathyris planosulcata | 67 | | | | 67 | |
| P. xungmiaoensis | 569 | 4 | | | 573 | |
| P. sp. | 1 | | | | 1 | 42 |
| Protathyris totals | | | | | 641 | |
| Atrypoidea inflata | 6 | | | | 6 | 0 |
| Howellella tingi | 15 | | | | 15 | 1 |
| Lingula orientalis | 19 | | | | 19 | 1 |
| Spirinella sp.? | 2 | | | | 2 | 0 |
| Eoschizophoria hesta | 29 | 13 | 7 | 2 | 51 | 3 |
| Total | | | | | 1,526 | 99 |
| **Locality ADH 33, Kuanti Formation** | | | | | | |
| Protathyrisina uniplicata | 5 | 33 | 44 | 18 | 100 | 100 |

*Environment:* Most of the shells occur as articulated specimens in a gray, calcareous mudstone. The medium diversity fauna, plus the occurrence of *Gypidula,* is consistent with a Benthic Assemblage 3 position, in a quiet-water environment as shown by the condition of articulation.

*Community Group: Striispirifer.*

### *Protathyrisina plicata, P. minor,* and *P. uniplicata* Community (Fig. 11; Table 15)

*Name:* This community is named here.
*Composition:* Dominant *Protathyrisina* with variable, lower numbers of *Protathyris, Eoschizophoria, Tadschikia, Aesopomum,* and *Howellella.*
*Age:* Late Ludlovian–Early Pridolian (see Wang and others, 1980, for conodont data).
*Typical Locality:* Qujing, eastern Yunnan.
*Geographic Distribution:* Eastern Yunnan in the Miaokao Formation, southern Yunnan (Yuanjiang County; unpublished material), and western Sichuan (Erlangshan; unpublished material).
*Environment:* The articulated nature of the shells, the dominance of the name-bearing genus, and the calcareous mudstone matrix, together with the nature of the minor constituents all suggest a restrictive, Benthic Assemblage 2 or 3 assignment. Low oxygen might be indicated, as well as little current activity.

The distinction between this community and the interlayered *Protathyris* Community might be due to the vagaries of larval settling and distribution rather than to any basic environmental differences.

*Community Group: Striispirifer.*

### *Protathyris–Lanceomyonia* Community (Fig. 13)

*Name:* This community is named here.
*Age:* The Middle Member of the Putonggou Formation occurs above the *Icriodus woschmidti*–bearing Lower Member (Qin and Gan, 1976), and below the *Spirigerina* fauna–bearing Upper Member. This suggests that the Middle Member is of Late Gedinnian age.
*Composition:* Qin and Gan (1976) cited *Protathyris, Lanceomyonia,* and *Howellella* as the most abundant, dominant genera in the Middle Member. We have a sample of this material available for study that confirms their work.
*Typical Locality:* Xia Wunagou Section, Diebu County, southeastern Gansu.
*Geographic Distribution:* Same as above.
*Environment:* The Middle Member of the Xia Putonggou Formation consists of calcareous slate, reefy limestone, and dolomitic limestone. The fossils occur as articulated individuals. Elsewhere in this Member there are additional low-diversity, high-dominance associations (including *Protocortezorthis, Linguopugnoides, Schizophoria,* and "*Stropheodonta,*" Fu Li-pu, written communication, 1980, as well as a few other associations including *Protathyris, Linguopugnoides,* and *Howellella*). The shells commonly occur in masses, i.e., tabular bodies, according to Dr. Fu (written communication, 1980). The overall aspect suggests a relatively shallow-water, possibly Benthic Assemblage 2 or shallow 3 environment within the quiet-water category, that features some type of restrictive factor to account for the low diversity.

*Community Group: Striispirifer.*

### *Protathyrisina uniplicata–Striispirifer* Community (Fig. 11; Table 16)

*Name:* This community is named here.
*Composition:* Abundant *Protathyrisina uniplicata* and *Striispirifer* (Table 16). Very low-diversity varying between dominant *Striispirifer* and divided dominance of *Striispirifer* and *Protathyrisina.*
*Age:* Ludlovian (Wang and others, 1980).
*Typical Locality:* Kuanti Formation, near Qujing, eastern Yunnan Province.
*Geographic Distribution:* Eastern Yunnan only, in the Kuanti Formation.
*Environment:* The shells occur in calcareous mudstone as disarticulated individuals in thin shell beds. The occurrence of abundant *Striispirifer* is consistent with a Benthic Assemblage 3 assignment, but the restrictive factor(s) responsible for the low diversity are uncertain. However, the disarticulated condition of the valves indicates a more turbulent environment than the *Atrypoidea-Protathyrisina uniplicata-Striispirifer* Community with which it is interbedded.

*Community Group: Striispirifer.*

### *Protochonetes–Rostrospirifer* Community (Fig. 15; Table 17)

*Name:* This community is named here.
*Age:* The community occurs in the Shizhou Member of the Yukiang Formation. This member is of Early Emsian age (Ruan and others, 1979; Wang Cheng-yuan and others, 1979).
*Composition:* Table 17 indicates the composition of a typical sample. Disarticulated pedicle valves of *Protochonetes* and *Rostrospirifer* dominate, with smaller numbers of *Dicoelostrophia, Howellella, Glyptospirifer,* and a stropheodontid. An allied collection (see Table 17) lacked specimens of *Protochonetes* and includes a specimen of *Athyris.* This indicates that the generic composition may vary, particularly when dealing with small samples (the second sample includes only 41 shells).
*Typical Locality:* Liujing, Hengxian County, southern Guangxi.
*Geographic Distribution:* Southern Guangxi, and possibly northern Sichuan if some of the material discussed by Chen (1979) belongs here.
*Environment:* The shells occur chiefly as disarticulated specimens in a calcareous mudstone matrix. If abundant protochonetids indicate turbid water, a fairly turbid environment in which there was a moderate level of current activity near the Benthic

TABLE 16. PROTATHYRISINA UNIPLICATA-STRIISPIRIFER COMMUNITY

| Name | Articulated Shell | Pedicle Valve | Brachial Valve | Either Valve | Total No. Specimens | % |
|---|---|---|---|---|---|---|
| **Locality ADH 27, Kuanti Formation** | | | | | | |
| Protathyrisina uniplicata | | 14 | 8 | 14 | 36 | 21 |
| Striispirifer yunnanensis | | 110 | 20 | 4 | 134 | 79 |
| Total | | | | | 170 | |
| **Locality ADH 29, Kuanti Formation** | | | | | | |
| Protathyrisina uniplicata | | 22 | 34 | 22 | 78 | 47 |
| Striispirifer yunnanensis | | 67 | 20 | 2 | 89 | 53 |
| Total | | | | | 167 | |
| **Locality ADH 30, Kuanti Formation** | | | | | | |
| Protathyrisina uniplicata | 7 | 2 | 2 | 1 | 12 | 52 |
| Striispirifer yunnanensis | | 10 | 1 | | 11 | 48 |
| Total | | | | | 23 | |
| **Locality ADH 30, Kuanti Formation (Silicified)** | | | | | | |
| Atrypoidea sp. | | 4 | 1 | | 5 | |
| Protathyrisina uniplicata | 235 | 332 | 352 | 4 | 923 | 46 |
| Striispirifer yunnanensis fragments | 3 | 280 | 397 | | 680 | 34 |
| **Locality ADH 27, Kuanti Formation** | | | | | | |
| Protathyrisina uniplicata | | 72 | 58 | 62 | 192 | 10 |
| Striispirifer yunnanensis | | 110 | 81 | 22 | 213 | 11 |
| Total | | | | | 2,013 | |

TABLE 17. PROTOCHONETES-ROSTROSPIRIFER COMMUNITY

| Name | Articulated Shell | Pedicle Valve | Brachial Valve | Either Valve | Total No. Specimens | % |
|---|---|---|---|---|---|---|
| **Locality ADH 10$_S$ (Locality I)** | | | | | | |
| Rostrospirifer aff. tonkinensis | 2 | 78 | 7 | | 87 | 30 |
| Dicoelostrophia spp. | 2 | 12 | 3 | | 17 | 6 |
| Howellella sp. | | 21 | 2 | | 23 | 8 |
| Glyptospirifer sp. | | 4 | 5 | | 9 | 3 |
| Protochonetes sp. | 7 | 120 | 25 | | 152 | 52 |
| Stropheodontid | | 2 | | | 2 | 1 |
| Total | | | | | 290 | |
| **Locality ADH 10 (Locality II)** | | | | | | |
| Rostrospirifer aff. tonkinensis | 25 | 14 | | | 39 | 95 |
| Athyris grandis | 1 | | | | 1 | 2 |
| Dicoelostrophia sp. | 1 | | | | 1 | 2 |
| Total | | | | | 41 | |

TABLE 18. <u>RETICULARIOPSIS</u>-"CHONETES" N. SP. COMMUNITY

| Name | Total No. Specimens | % |
|---|---|---|
| <u>Reticulariopsis</u> cf. <u>eifeliensis</u> | 147 | 24 |
| "Chonetes" n. sp. | 92 | 15 |
| <u>Levenea</u> n. sp | 78 | 13 |
| <u>Spinatrypa</u> n. sp. | 74 | 12 |
| <u>Kayseria</u> <u>lens</u> | 47 | 8 |
| <u>Douvillina</u> cf. <u>corrugata</u> | 37 | 6 |
| <u>Athyris</u> sp. | 20 | 3 |
| <u>Aulacella</u> <u>eifeliensis</u> | 16 | 3 |
| <u>Plectospira</u> n. sp. | 10 | 2 |
| <u>Minutostropheodonta</u> n. sp. | 11 | 2 |
| <u>Cyrtina</u> <u>heteroclyta</u> | 15 | 2 |
| <u>Spinocyrtia</u> <u>martinaofi</u> | 8 | 1 |
| <u>Undispirifer</u> sp. | 9 | 2 |
| <u>Skenidium</u> <u>polonicum</u> | 8 | 1 |
| <u>Teichertina</u> cf. <u>americana</u> | 9 | 2 |
| <u>Malurostrophia</u> sp. | 6 | 1 |
| <u>Nadiastrophia</u> sp. | 2 | - |
| <u>Leptostrophia</u> sp. | 7 | 1 |
| <u>L.</u> n. sp. | 2 | - |
| <u>Schellwienella</u> sp | 2 | - |
| <u>Gypidula</u> sp. | 4 | 1 |
| <u>Antirhynchonella</u> sp. | 5 | 1 |
| <u>Uncinulus</u> cf. <u>parallelopipedus</u> | 4 | 1 |
| <u>Spirigerina</u> sp. | 1 | - |
| <u>Parathyrisina</u> sp. | 1 | - |
| ?<u>Striispirifer</u> sp. | 2 | |
| Total | 617 | |

Assemblage 2-3 boundary, is indicated. The shells may be chiefly Benthic Assemblage 2-3 eurytopic types.

Consideration of Table 17 suggests that some ecologically significant transport of the dominant taxa might have taken place. For example, the possibility that the *Protochonetes* were transported into the depositional site cannot be discounted with the data available to us.

***Community Group:*** *Striispirifer.*

### Receptaculitid Community (Fig. 15)

***Name:*** This community is named here.

***Age:*** Although not very richly fossiliferous, the Daliancum Member of the Yukiang Formation is of Early Emsian age because of its stratigraphic position (Ruan and others, 1979; Wang Cheng-yuan and others, 1979).

***Composition:*** The unit consists of a single bedding plane covered with in situ receptaculitids. However, the fossils have not been studied, so their generic and specific identity is uncertain, although they are clearly receptaculitids.

***Typical Locality:*** None designated.

***Geographic Distribution:*** Receptaculitid-dominated beds are known in the Ordovician through Devonian (Clarke, 1922, p. 239–240; Nitecki, 1972, p. 84–88).

***Environment:*** The muddy matrix suggests a relatively quiet water environment. The absence of a rich, shelly fauna suggests that receptaculitid communities reflect some sort of specialized conditions. This is probably a Benthic Assemblage 3 unit because

of its position between Benthic Assemblage 3 units, and because receptaculitids elsewhere commonly occur in this position. If receptaculitids are calcareous algae, as many now conclude, then a photic zone assignment is required.

***Community Group:*** We are not in a position to suggest a useful community group terminology for the receptaculitids.

### Reticulariopsis–"Chonetes" n. sp. Community (Fig. 17, Table 18)

***Name:*** This community is named here.

***Age:*** Zhang Yan (written communication, 1980) assigns the Yikewusu Formation to the Eifelian, as is also evident from her faunal list (Table 18).

***Composition:*** Dominant *Reticulariopsis,* plus abundant *"Chonetes"* n. sp., and several other genera, and a number of rarer genera to give rise to a high-diversity community (see Table 18).

***Typical Locality:*** Xipingshan, northwest Neimongol.

***Geographic Distribution:*** Same as above.

***Environment:*** The shells occur in limestone as dominantly articulated specimens. The overall high diversity and the presence of *Kayseria, Skenidium,* and *Teichertina* indicate a Benthic Assemblage 4-5 assignment, and a relatively normal level of current activity. This community unit is similar to some of the Eifelian and Givetian faunas described by Johnson (1970, 1978).

***Community Group:*** *Dicoelosia-Skenidioides.*

***Discussion:*** Below the Eifelian unit, Dr. Zhang Yan (written communication, 1980) reported a lower unit of massive limestone containing articulated specimens of *Uncinulus subcordi-*

TABLE 19. <u>ROSTROSPIRIFER-ATHYRISINA</u> COMMUNITY, COLLECTION 25

| Name | Articulated Shell | Pedicle Valve | Brachial Valve | Either Valve | Total No. Specimens | % |
|------|---------|---------|---------|---------|---------|---|
| Rostrospirifer | 371 | 36 | 7 | | 414 | 65 |
| Athyrisina | 112 | 1 | | | 113 | 18 |
| Glyptospirifer | 72 | 22 | | | 94 | 15 |
| Xenostrophia | 12 | | 1 | | 13 | 2 |
| Total | | | | | 634 | |

formis and *U. parallelopipedus.* Below this, the Late Emsian Zhusilen Formation calcareous sandstone contains disarticulated specimens of *"Hysterolites"* n. sp., *"H."* sp., *Atrypa,* and *"Chonetes."* Farther down are beds yielding *Fallaxispirifer pseudofallax* and *Fascistropheodonta sedgwicki,* together with *Parachonetes, Devonochonetes* and *Desquamatia,* plus *Paraspirifer.* Still further down are sandy conglomeratic beds containing partially disarticulated *Badaingarania striata, Glossinotoechia* n. sp., *Schizophoria* sp., *Rariella, Rhynchospirina, Megakozlowskiella, ?Fimbrispirifer, Pararhynchospirina,* and *Neimongolella,* with *Megaplectatrypa* and *Stegerhynchus.*

### *Rhipidothyris* Community (Fig. 17)

*Name:* This community was defined by Boucot, Massa and Perry (1983).
*Age:* Late Eifelian (Wang Cheng-yuan and others, 1979) based on stratigraphic position, and regional relations. The unit occurs in the upper part of the Yingtang Formation.
*Composition:* Moderately abundant *Rhipidothyris* and a less abundant spiriferid.
*Typical Locality:* South of Nandang, Maanshan, Xiangzhou County, central Guangxi.
*Geographic Distribution:* Same as above.
*Environment:* The fossils occur as articulated valves in a calcareous mudstone matrix. The low diversity suggests that the environment was not only quiet but also restricted. We suspect that this is a Benthic Assemblage 3, or possibly 2, community. This inference is based on the position of the *Globithyris* and *Rhenorensselaeria* Communities, genera closely related to *Rhipidothyris,* that are known in Benthic Assemblage 2 (Boucot, 1975).
*Community Group: Striispirifer.*
*Discussion:* Hou (1963) described *Rhipidothyris sulcatilis* as a punctate shell, but he found no loop as would be expected in a terebratuloid. Yue and Bai (1978) erected a punctate rhynchonellid genus, selected Hou's *R. sulcatilis* as the type species, and described an additional species of their new genus as *Yingtangella minor.* Wang and Zhu (1979) described two additional species of *Rhipidothyris* (*R. ovata* and *R. bicostata*), while also observing punctae, but failed to find a loop. The resulting situation yields a Late Eifelian, *Rhipidothyris*-like shell lacking a loop—which may

actually belong to *Rhipidothyris*—that has been recovered from central Guangxi.

### *Rhynchotrema chechiangensis* Community (Fig. 7)

*Name:* This community is named here.
*Composition:* About 90 percent of articulated specimens of *Rhynchotrema,* associated with a few trilobites (*Remopleurides* and a *Cheirurus*), gastropods, and about six other brachiopod genera.
*Age:* Chen Xu (oral communication, 1980) regarded this community (from the Xiazheng Formation) as of Late Ordovician age in a general sense. No precisely age-diagnostic fossils have yet been recognized.
*Typical Locality:* Xiazheng Bridge, Jiudu, Yushan County, northeastern Jiangxi.
*Geographic Distribution:* The border region between northeastern Jiangxi and adjacent Anhui Provinces.
*Environment:* The fossils occur in gray, calcareous shale, and blue-gray argillaceous, micritic limestone as articulated, relatively high-density aggregations. This is a quiet-water environment, of the thinly bedded, or laminated, low-oxygen type. The *R. chechiangensis* Community is similar to Boucot's (1975), Benthic Assemblage 1 "Rhynchonellid Community" of the Silurian-Devonian, as well as to similar Upper Ordovician communities described by Bretsky (1970) from the Appalachians. But the presence of the trilobites and the additional brachiopod taxa in the Jiangxi community indicate that a Benthic Assemblage 2 assignment is more likely.
*Community Group:* None selected. The rhynchonellid taxa involved in these rhynchonellid-rich communities need to be studied more carefully before a community group nomenclature can be reliably established.

### *Rostrospirifer–Athyrisina* Community (Fig. 16; Table 19)

*Name:* This community is named here.
*Age:* The Luhuei Member of the "Sipai Formation" is of Late Emsian or Dalejan age (Wang Yu and others, 1979; Wang Cheng-yuan and others, 1979), as indicated by the presence of *Nowakia cancellata* and *N. richteri, Polygnathus inversus,* and *P. serotinus,* plus *Anarcestes noeggerati.*

*Composition:* This is a medium-diversity unit (Table 19) with high dominance of three genera.

*Typical Locality:* Dale, Xiangzhou County, central Guangxi.

*Geographic Distribution:* Same as above.

*Environment:* The fine-grained matrix combined with the low diversity suggests some kind of restrictive factor. A soft bottom might be involved. This is a Benthic Assemblage 3 community. There is a high percentage of articulated valves.

*Community Group: Striispirifer.*

### *Septoparmella* Containing *Dicoelosia–Skenidioides* Community Group (Fig. 13)

*Name:* See discussion of *Dicoelosia-Skenidioides* Community Group. Notanopliids occur in this Community Group in some areas during the Lower Devonian (see Boucot, 1975).

*Age:* See discussion of *Dicoelosia-Skenidioides* Community Group and *Striispirifer-Molongia* Community.

*Composition:* Liu and Huang (1977) report abundant *Idioglyptus,* fairly abundant *Isorthis,* and less abundant *Howellella, Salopina, Septoparmella* (their *"Metaplasia"*), and *Leptostrophia.*

*Geographic Location:* Erdaogou, Yongji County, Jilin Province.

*Typical Locality:* None designated.

*Environment:* Liu and Huang report that the upper part of the Erdaogou Formation containing this unit consists of limestone and siltstone. The illustrations of the brachiopods are of casts and molds of disarticulated specimens in the siltstone. The presence of a notanopliid indicates a Benthic Assemblage 4-5 assignment. This is presumably a normal diversity outer shelf community, but additional data are needed before more is known about its affinities. Somewhat similar communities (*Maoristrophia* Community) are mentioned by Boucot (1975) from the Lower Devonian of Australia. It should be noted that notanopliids may occur either by themselves as low-diversity, high-dominance communities, or as low-abundance members of high-diversity communities—the latter is what the Erdaogou Formation represents.

*Community Group: Dicoelosia-Skenidiodes.*

### *Septoparmella–Aldanispirifer* Community (Fig. 13)

*Name:* This community is named here.

*Composition:* Su Yang-Zheng (written communication, 1980) reported scattered, disarticulated specimens of *Septoparmella* and *Aldanispirifer* associated with conglomerate in argillaceous mudstone.

*Age:* Dr. Su (written communication, 1980) reported this community from the lower part of the Barantanhua Formation of probable Siegenian age; a Gedinnian age cannot be ruled out on the basis of the available evidence.

*Typical Locality:* None designated.

*Geographic Distribution:* Dong Ujimqin Qi County, Neimongol.

*Environment:* This low-diversity community, occurring as scattered shells, suggests a relatively quiet-water, possibly Benthic Assemblage 4-5, environment of the kind where notanopliids most commonly are relatively dominant or important, although not enough evidence is available for us to be certain.

*Community Group: Dicoelosia–Skenidioides.*

### *Septoparmella* Community (Fig. 13)

*Name:* This community is named here.

*Age:* The unpublished South China occurrence, from an unnamed formation, is assigned to the Gedinnian because of the presence of *Septoparmella.*

*Composition:* The Guangxi Province occurrence contains *Septoparmella, Stegerhynchus?* sp., *Isorthis?* sp., *Leptostrophia?* sp., and cf. *Metaplasia? rectilateralis* Borisiak (=*Septoparmella*) as scattered, disarticulated valves in a calcareous mudstone matrix.

*Typical Locality:* None designated because of the very preliminary nature of our information.

*Geographic Distribution:* Lingshan, southern Guangxi.

*Environment:* This is a low-diversity, high-dominance, low-population density community that has much in common with notanopliid communities known from Australia, as well as from the Rhineland, United States (Nevada), North China, and elsewhere, including younger beds in South China. Presumably, it represents a similar deep-water, relatively quiet-water, Benthic Assemblage 4-5, environment. The restrictive factor(s) are unknown.

*Community Group:* The notanopliid subgroup branch of the *Dicoelosia-Skenidioides* Community Group.

### *Shaleriid–"Dalmanella"* Community (Fig. 9)

*Name:* This community is named here.

*Composition:* Abundant shells of an unidentified shaleriid (no brachial valves present in the available collection), *"Dalmanella,"* aff. *Anabaria,* and a few specimens of *Skenidioides* sp.

*Age:* Yan Guo-shun (oral communication, 1980) reported the presence of the Middle Llandoverian–age graptolites *Rastrites* sp. and *Demirastrites triangularis* 200 m above this unit (see Mu and others, 1986, for the age of local rock units).

*Typical Locality:* Near Wangguangou, Xichuan County, Henan, in the easternmost Qinling Mountains.

*Geographic Distribution:* Same as above.

*Environment:* The thoroughly disarticulated condition of the brachiopods suggests a moderate amount of current activity. The presence of *Skenidioides* indicates a Benthic Assemblage 4-5 assignment.

*Community Group:* The presence of *Skenidioides* indicates a *Dicoelosia-Skenidioides* Community Group assignment. The relatively small size of the available collection does not rule out the possibility that additional genera, including *Dicoelosia* will be found here.

TABLE 20. <u>SINOSTROPHIA-DISCOMYORTHIS</u> COMMUNITY*

| Name | Total No. Specimens | % | Page Ref. |
|---|---|---|---|
| "<u>Lingula</u>" | 1 | – | 37 |
| <u>Schizophoria</u> sp. indet. | 1 | – | 37 |
| <u>Proschizophoria</u> <u>kobayashi</u> | 6 | 2 | 38 |
| <u>Discomyorthis</u> <u>kinsuiensis</u> | 40 | 15 | 41 |
| <u>Reeftonia</u> <u>borealis</u> | 20 | 8 | 43 |
| <u>Leptaenopyxis</u> <u>bouei</u> | 3 | 1 | 46 |
| "<u>Megastrophia</u>" (<u>Protomegastrophia</u>) <u>manchurica</u> | 6 | 2 | 47 |
| <u>Leptostrophia</u> <u>nonakai</u> | 10 | 4 | 48 |
| <u>Sinostrophia</u> <u>kondoi</u> | 70 | 27 | 53 |
| <u>Pholidostrophia</u> sp. indet. A | 2 | 1 | 57 |
| <u>Pholidostrophia</u> sp. indet. B | 2 | 1 | 57 |
| "<u>Schuchertella</u>" sp. | 9 | 3 | 58 |
| <u>Aesopomum</u> <u>chinense</u> | 12 | 5 | 59 |
| <u>Chonostrophiella</u> <u>khinganensis</u> | 12 | 5 | 61 |
| <u>Zlichorhynchus</u> <u>asiaticus</u> | 3 | 1 | 63 |
| <u>Wilsoniella</u> <u>grandis</u> | 8 | 3 | 65 |
| <u>Uncinulus</u> <u>piloides</u> | 2 | 1 | 67 |
| <u>Pectorhyncha</u>? sp. | 3 | 1 | 68 |
| <u>Eucharitina</u>? sp. | 2 | 1 | 68 |
| "<u>Camarotoechia</u>" sp. | 2 | 1 | 69 |
| <u>Atrypa</u> sp. | 1 | – | 70 |
| <u>Coelospirella</u> <u>orientalis</u> | 4 | 2 | 72 |
| <u>Nucleospira</u> <u>musculosa</u> | 4 | 2 | 73 |
| <u>Cyrtina</u> sp. | 3 | 1 | 74 |
| <u>Acrospirifer</u> sp. | 5 | 2 | 74 |
| "<u>Howellella</u>" <u>amurensis</u> | 25 | 10 | 76 |
| <u>Paraspirifer</u> aff. <u>cultrijugatus</u> | 6 | 2 | 78 |
| "<u>Fimbrispirifer</u>" cf. "<u>divaricatus</u>" | 1 | – | 80 |
| Total | 263 | | |

*From Hamada, T., 1971, Palaeontological Society of Japan Special Paper 15.

*Sinostrophia–Discomyorthis* Community (Table 20)

*Name:* This community is named here.

*Age:* Hamada (1971) favored an Early Emsian age for the community. His discussion suggests that this is the most likely assignment based on known ranges of the genera he recognized. It is unlikely that the age will be outside the Late Siegenian-Emsian range.

*Composition:* Hamada (1971) illustrated the unit very well, while discussing its nature, and Modzalevskaya (1968) illustrated some of its elements from a nearby Soviet Union locality. Hamada's Table 5 provides a complete list of the fauna with a good idea of relative abundances (see our Table 20). The shell assigned to *Fimbrispirifer* cannot be shown to definitely belong to that genus. Similar spiriferids in Asia also possess plications in the fold and sulcus, and the absence of proved *Fimbrispirifer* outside of the Eastern Americas Realm makes his determination unlikely. Hamada's *Dalejina kinsuiensis* is now assigned to *Discomyorthis*, and his *Bifida* is now assigned to *Coelospirella*.

*Typical Locality:* Houlungmen region of the Lesser Khingan Mountains.

*Geographic Distribution:* Northeast China, Neimongol, and adjacent parts of the Soviet Union.

*Environment:* This is a typical Benthic Assemblage 3 community. The shells occur in a calcareous, argillaceous, medium- to fine-grained greenish sandstone, as largely disarticulated specimens. The size of the faunal list (Hamada, 1971, Table 5), as well as the relative rarity of some of the shells, suggests that a few of the rarer taxa may have become admixed by current transport, but present data are inadequate to decide this question.

Assemblages of this type normally represent relatively normal, level-bottom conditions within the Benthic Assemblage 3 region.

*Community Group: Striispirifer.*

*Spirigerina* Community (Fig. 10)

*Name:* This community is named here.

*Composition:* The composition of this unit is variable, but is always very low diversity. It may include dominant *Spirigerina*, slightly dominant *Striispirifer* (of the type with U-shaped interspaces that has been referred to elsewhere as *Hedeina*), and a smaller percentage of *Nalivkinia* and *Zygospiraella*.

*Age:* Rong and Yang (1981) placed this community near the Middle-Upper Llandoverian boundary.

*Typical Locality:* Juntianba, Baisha, Shiqian County, northeastern Guizhou.

*Geographic Distribution:* Leijiatun, Shiqian county, northeastern Guizhou Province, also at Juntianba, Baisha, Shiqian County, and a few other localities near the two cited here.

*Environment:* Calcareous, gray mudstone occupying an interval of about 10 m yields this fauna. Most of the shells are articulated and well sorted. This community is found in the middle part of the Xiangshuyuan Formation. The community appears to be a low- to medium-diversity Benthic Assemblage 3 unit, intermediate in environmental terms between the *Eospirigerina, Nalivkinia,* and *Zygospiraella-Brachyelasma* Communities of the lower Silurian. Because an abundance of *Eospirigerina* is followed in time by a similar abundance of *Spirigerina,* the same community group is certainly indicated. The precise factors that are responsible for the slight differences in composition between this community and the closely related ones are uncertain.

*Community Group: Striispirifer.*

## *Stricklandia–Merciella* Community (Fig. 10)

*Name:* This community was named by Rong and Yang (1981). The dominance of *Brevilamnulella* over *Merciella* is overlooked in naming in the interests of nomenclatorial stability.

*Composition:* Medium-diversity brachiopod-dominated community with about one-third *Stricklandia transversa* plus *Brevilamnulella, Isorthis, Atrypina, Pleurodium,* and *Beitaia. Merciella* may also be present. *Brevilamnulella* is second in abundance.

*Age:* Lower part of the Upper Llandoverian (Rong and Yang, 1981).

*Typical Locality:* Heshui, Yinjiang County, northeastern Guizhou.

*Geographic Distribution:* More than four localities are known in southern China.

*Environment:* The relatively high abundance of both *Stricklandia* and *Brevilamnulella* suggests a position near the Benthic Assemblage 4-5 boundary. The absence of such taxa as *Dicoelosia* and *Skenidioides* indicates rougher water than is normal, although not strong enough to accomplish much disarticulation of the brachiopods. The rock type is a gray, calcareous mudstone.

*Community Group:* A mixture of Stricklandiidae, Virgianidae, and *Dicoelosia-Skenidioides* Community Groups. Boucot (1975, p. 262) used the term Undivided Communities for a high diversity, *Dicoelosia-Skenidioides* Community Group unit present in the Benthic Assemblage 5 position. On further consideration, it would be best to designate this the *Brevilamnulella-Cyrtia* Community, since either one or both of these genera (or of the descendant genus *Clorinda*) are commonly fairly abundant. Both *Brevilamnulella* and *Cyrtia,* as well as *Clorinda,* may occur in Benthic Assemblage 4, high-diversity *Dicoelosia-Skenidioides* Community Group units, but not in abundance.

*Discussion:* Rong and Yang (1981) named the community with the abundance of *Stricklandia* and the relative rarity of *Merciella* in mind.

## Low Diversity *Striispirifer* Community (Fig. 10)

*Name:* This community is named here.

*Composition:* A low-diversity, high-dominance community featuring abundant *Striispirifer* and a few bivalves.

*Age:* Upper Llandoverian (Ge and others, 1979).

*Typical Locality:* Leijiatun, Shiqian County, Guizhou, in the Xiushan Formation.

*Geographic Distribution:* Northeastern Guizhou.

*Environment:* Communities rich in *Striispirifer,* including some with the *Hedeina* form, are abundant in the later Llandoverian of South China. They appear to belong in Benthic Assemblage 3, probably in the shallow portion. Above the Leijiatun occurrence of the Low Diversity *Striispirifer* Community, an occurrence of the *Striispirifer-Nalivkinia* Community is assigned to Benthic Assemblage 3; and underlying, an occurrence of the *Nucleospira-Nalivkinia* Community is also assigned to Benthic Assemblage 3. The presence of disarticulated shells in a mudstone matrix is consistent with some type of restricted environment, i.e., low diversity and relatively quiet water.

This community contrasts markedly with the normal, high-diversity *Striispirifer* Community (see Boucot, 1975).

*Community Group: Striispirifer.*

## *Striispirifer–Molongia* Community (Fig. 12)

*Name:* This community is named here.

*Composition:* Liu and Huang (1977) indicate that *Striispirifer* is the most abundant genus, followed by *Molongia, Aegiria,* and *Iridostrophia* in lower abundance, and smaller numbers of *Eospirifer, Eoreticularia, Proreticularia, Nikiforovaena, Salopina,* and *Amphistrophia.* We have not, however, had access to actual specimens. Corals and trilobites are also present.

*Age:* This community was described and figured for the first time from the Zhangjiatun Formation by Liu and Huang (1977). We consider the age of the fauna as most likely Ludlovian. It occurs below the Erdaogou Formation that contains the notanopliid *Septoparmella* (Liu and Huang's *"Metaplasia"*) in its upper part. Notanopliids are unknown elsewhere beneath the Gedinnian. The presence in the lower part of the Erdaogou Formation of *Protathyrisina,* a genus known elsewhere no higher than the Pridolian is also consistent with the conclusion that the *Molongia*-bearing, underlying Zhangjiatun Formation is of Ludlovian age. However, the bulk of the fauna consists of genera that could range throughout the Ludlovian-Gedinnian time interval.

*Typical Locality:* Zhangjiatun, Yongji County, Jilin Province.

*Geographic Distribution:* Same as above.

*Environment:* The Zhangjiatun Formation, according to Liu and Huang (1977), consists of fine-grained sandstone, siltstone, and mudstone. Their illustrations indicate that the disarticulated brachiopods occur as casts and molds in clastic rocks. This community is clearly related to the *Striispirifer niagarensis* Community whose typical locality is in the Rochester Shale (Boucot, 1975). The *Striispirifer-Molongia* Community can be viewed as an endemic community belonging to the *Striispirifer* Community Group. A Benthic Assemblage 3, relatively normal, quiet-water environment is indicated for this relatively high-

TABLE 21. STRIISPIRIFER-NALIVKINIA COMMUNITY

| Name | Articulated Shell | Pedicle Valve | Brachial Valve | Either Valve | Total No. Specimens | % |
|------|------|------|------|------|------|------|
| <u>Striispirifer</u> | | 110 | 56 | 79 | 245 | 65 |
| <u>Nalivkinia</u> | | 14 | 26 | 77 | 117 | 31 |
| <u>Linostrophomena</u> | | 6 | 3 | 7 | 16 | 4 |
| Total | | | | | 378 | |

| | Pygidium | Cephalon | | | | |
|------|------|------|------|------|------|------|
| <u>Coronocephalus</u> | 4 | 5 | | | 9 | |

diversity community, with enough current activity to have disarticulated the shells.

***Community Group:*** *Striispirifer.*

### *Striispirifer–Nalivkinia* Community (Fig. 10; Table 21)

***Name:*** This community is named here.

***Composition:*** A low- to medium-diversity community featuring dominant *Striispirifer* and large *Nalivkinia.* Table 21 indicates the generic composition of a typical sample.

***Age:*** Upper Llandoverian; *celloni* Zone conodonts closely associated below. Ge and others (1979) assigned the bed containing this community to the Wenlockian, but the close association with the underlying bed yielding *celloni* Zone conodonts makes an Upper Llandoverian age more likely.

***Typical Locality:*** Leijiatun, Shiqian County, Guizhou.

***Geographic Distribution:*** Northeastern Guizhou.

***Environment:*** At the typical locality, the fauna occurs in a shell bed with a mud-size, calcareous matrix. The shells are thoroughly disarticulated. The overall generic compositions of the fauna is consistent with a Benthic Assemblage 3, relatively quiet-water environment.

***Community Group:*** *Striispirifer.*

### *Stringocephalus* Community

***Name:*** Boucot (1975) briefly discussed the *Stringocephalus* Community as a low-diversity community dominated by the name bearer.

***Age:*** In South China it occurs in the Tongkangling Formation of Givetian age (Wang and Yu, 1962; Wang and others, 1974; Wang Yu and others, 1979).

***Composition:*** The sample available to us from the Liujing area consists almost entirely of *Stringocephalus* with minor numbers of *Bornhardtina* and *Rensselandia.*

***Typical Locality:*** None designated.

***Geographic Distribution:*** Worldwide outside of the Malvinokaffric Realm (see Boucot and others, 1966, for a good example

of the localities; most localities yielding *Stringocephalus* belong to the *Stringocephalus* Community).

***Environment:*** The Liujing occurrence of the community is unusual in that the shells are scattered and occur chiefly as articulated, unsorted individuals. We must conclude from the articulated nature of the shells, the wide size range, and the micritic, thin-bedded, dark-colored nature of the entombing limestone that this represents a Benthic Assemblage 3, quiet-water, rather than the normal rough-water environment occurrence. The *Stringocephalus* beds at Liujing intertongue with platy *Nowakia* Community-bearing strata that are also of quiet water type. *Stringocephalus* possibly developed as spat adjacent to rougher water, small bioherms (also present in the Tongkangling Formation), and sometime later in life were swept into this final quiet-water environment. However, the spat possibly settled where they were finally entombed, although this is most unusual for the genus. *Stringocephalus* commonly occurs in a disarticulated condition.

***Community Group:*** Boucot (1975) did not consider the question of community group assignment for the *Stringocephalus* Community. The rough-water, Middle Devonian communities dominated by *Rensselandia,* as well as descendent, closely allied taxa such as *Stringocephalus* and allied genera, form a natural unit. We propose the term *Rensselandia* Community Group for these communities.

### *"Subcuspidella"–Athyrisina* Community (Fig. 16; Table 22)

***Name:*** This community is named here.

***Age:*** The Shipeng Member of the "Sipai Formation" is of Late Emsian or Dalejan age (Wang Yu and others, 1979; Wang Cheng-yuan and others, 1979), as indicated by the presence of *Polygnathus inversus, Nowakia cancellata,* and *Anarcestes noeggerati.*

***Composition:*** This is a low-diversity community (Table 22) containing *"Subcuspidella"* and *Athyrisina* only.

***Typical Locality:*** Dale, Xiangzhou County, central Guangxi.

***Geographic Distribution:*** Same as above.

*Environment:* The wholly articulated shells occur in a micritic matrix. The relatively long pedicle interarea and large deltidial structures of *"Subcuspidella"* are consistent with life on a soft mud substrate that excluded most other benthonic animals. Many spiriferids and spiriferid-like shells with an elongate pedicle interarea can be interpreted as forms modified for life on a soft substrate (see also the *Cyrtia* Community in the Silurian, and Boucot, 1984, for a brief discussion and references).
*Community Group:* Striispirifer.

## Trimerella Community (Fig. 10)

*Name:* This community is used in the same sense as Boucot (1975).
*Composition: Trimerella* only.
*Age:* The beds yielding this community at Juntianba are probably of very early Upper Llandoverian age (Rong and Yang, 1981), as is the Hanjiadian area material.
*Typical Locality:* None designated.
*Geographic Distribution:* Juntianba and Hanjiadian area of northern Guizhou.
*Environment:* Although trimerelloids normally lived in a relatively rough-water environment, they disarticulated with difficulty. The specimens from Juntianba are large articulated shells in a massive limestone, whereas those from Hanjiadian are disarticulated, although also present in a massive limestone. A Benthic Assemblage 3 assignment is made, and further confirmed by the presence at Juntianba of an overlying bed of corals, which in turn is overlain by a bed of algal oncolites, and that is overlain by the Benthic Assemblage 3 *Paraconchidium–Virgianella* Community.
*Community Group:* Although Boucot (1975, p. 236) did not erect a community group for trimerelloid-dominated units, it is now apparent that there is a need for a Trimerelloid Community Group to include units dominated by trimerelloid genera such as *Trimerella* and *Dinobolus.* We here propose the community group name.

## Trimurellina–"Eoconchidium" Community (Fig. 7)

*Name:* This community is named here.
*Composition:* A high-diversity community containing more than 10 genera. The collections from Zhejiang and Anhui Provinces are too small to provide a reliable estimate of total diversity, but we would estimate that about 20 genera will ultimately be found in this community. The fauna available to us consists of *Trimurellina,* "Eoconchidium" (a form with a marked fold and sulcus), *Foliomena, Christiania, Sowerbyella, Eospirigerina, Kassinella, Cyclospira,* and at least four other distinctive forms that have not yet been identified satisfactorily.
*Age:* The composition of the fauna indicates a probable early Ashgillian age. The Huangnigang Formation rests below the Wufeng Formation, graptolite-bearing equivalent, the Yuqian Formation of the Ashgillian; it also lies above a Caradocian, Pagoda Limestone equivalent. The presence of the *Trimurellina–"Eocon-*

TABLE 22. "SUBCUSPIDELLA"-ATHYRISINA COMMUNITY

| Name | Articulated Shell | % |
|---|---|---|
| Collection 15 | | |
| "Subcuspidella" | 251 | 93 |
| Athyrisina | 20 | 7 |
| Total | 271 | |

*chidium"* Community is consistent with an Ashgillian age, particularly in view of the presence of a virgianid brachiopod.
*Typical Locality:* South foot of Jiangshan, just north of Jiangshan County town, western Zhejiang Province.
*Geographic Distribution:* Western Zhejiang Province, also Anhui Province.
*Environment:* The shells occur in a calcareous, very fine-grained mudstone as scattered, disarticulated shells. The absence of dicoelosids or *Skenidioides* suggests, in the presence of a high number of plicate virgianids, that this is an assemblage representing a normal marine environment near the Benthic Assemblage 3-4 boundary. However, P. M. Sheehan (written communication, 1982) concluded from the presence of *Foliomena* and *Cyclospira* that a deeper position is more likely.

We would not ordinarily propose a community based on such limited material, but because of its environmental and biogeographic importance, we have done so.
*Community Group:* We cannot yet provide a community group assignment for this unit due to overall ignorance of how best to assign the Ordovician faunas.

## Tuvaella gigantea Community (Fig. 12)

*Name:* Used in the sense of Boucot (1975).
*Composition:* Predominantly *T. gigantea* with a considerable admixture of *Leptostrophia* (brachial valves unknown here, but the stratigraphic position suggests this generic assignment as *Protoleptostrophia* has not been found in beds earlier than Siegenian), *Tannuspirifer, Meristina, Isorthis, Leptaena "rhomboidalis," Stegerhynchus,* and *Strophonella.* See Rong and Zhang (1982) for details.
*Age:* Ludlovian in age, following Vladimirskaya (1972; Vladimirskaya and Chekhovich, 1969) and Rong and Zhang (1982).
*Typical Locality:* None designated.
*Geographic Location:* Barkol, Xinjiang.
*Environment:* The disarticulated nature of the shells, except for *Meristina,* and the occurrence in a fine- to medium-grained, argillaceous sandstone, suggest a moderate amount of current movement. The overall aspect of the fauna suggests a Benthic Assemblage 3 assignment. Boucot (1975) indicated a Benthic Assemblage 2 location for the *Tuvaella* Community. However, the nature of the Xinjiang collection alone suggests a Benthic Assemblage 3 position. Jones (1980) reported *Tannuspirifer* in Arc-

tic Canada associated with a typical Benthic Assemblage 3 assemblage including *Gypidula*; this tends to support a Benthic Assemblage 3 assignment for *Tuvaella*. Su Yang-zheng (oral communication, 1980) provided the following data from an unnamed *Tuvaella*-bearing community on the right bank of the Guanniaohe River of the Lesser Hingan (Su's Locality IX, p. 29). An Early Devonian marine unit is underlain by another unit containing scattered linguloids. This unit, in turn, is underlain by a unit containing some beds with *Tuvaella* and *Meristina*. These *Tuvaella* beds are underlain by another unit yielding both *Dicoelosia* and *Janius* (Su's Locality IX, p. 29). In view of these data, we conclude that *Tuvaella* probably occurs in Benthic Assemblages 2 and 3.

*Community Group:* Boucot (1975, p. 236) implied that the *Tuvaella* Community could not be assigned to the community groups that he discussed. The Barkol data (Rong and Zhang, 1982), as well as information from other localities, demonstrate that this Community is no more than an endemic community belonging to the *Striispirifer* Community Group. Copper (1977a) derived *Tuvaella* from a cosmopolitan zygospirid ancestor; he concluded that the Benthic Assemblage position of the zygospirids in the Ordovician is similar to what we have concluded for *Tuvaella*.

## *Vagrania–Leptathyris* Community (Fig. 15)

*Name:* This community is named here.

*Age:* Early Emsian (see Wang Cheng-yuan and others, 1979) for occurrences of this unit that are dated by the presence of *Nowakia barrandei*.

*Composition:* Although this community is currently known in South China from only a single small collection, its importance requires that it be considered. The relatively normal diversity fauna contains the following taxa: *Vagrania, Leptathyris, Pentamerella, Carinatina,* chonetid, *Brevispirifer?, Desquamatia?,* fine-ribbed *Gypidula?, Spinatrypa, Anatrypa, Quadrithyris?,* bilobate rhynchonellid, *Atrypa* 1, *Atrypa* 2, *Nymphorhynchia, Schizophoria?, Uncinulus,* and a leiorhynchid.

*Typical Locality:* Daliancun, Nanning area, southern Guangxi.

*Geographic Distribution:* Same as above.

*Environment:* The fossils occur largely as articulated valves in a siliceous siltstone matrix indicating relatively quiet water. Although the community is not well located environmentally in South China, except that it is near the outer limits of the shelly facies–Nandan facies, it may be compared to similar units elsewhere. The Eifelian-age *Leptathyris* faunas of Nevada (see Johnson, 1970, for examples) have much in common ecologically with this community—particularly the higher diversity *Leptathyris* communities. The community group appropriate for the Nevada material appears earlier in South China. Although our sample size is small, *Leptathyris* and *Vagrania* are the most abundant shells present. This community belongs in about the Benthic Assemblage 3-4 position, both in South China and in

Nevada. This community comes from an unnamed formation in South China.

*Community Group: Striispirifer.*

## *Zdimir* Community (Figs. 16 and 17)

*Name:* Boucot (1975) briefly used this name for one of the many gypidulid communities. We use it here in the same sense.

*Age:* Late Emsian and Early Eifelian (Wang Yu and others, 1979; Wang Cheng-yuan and others, 1979) based on lateral relations with conodont-bearing and *Nowakia*-bearing beds, as well as with ammonite-bearing beds in China. Elsewhere in Eurasia this community is known well into the Givetian.

*Composition:* Chiefly disarticulated, well-sorted, large valves of *Zdimir* with a predominance of pedicle valves, as is normal with gypidulinids, in the shell coquina.

*Typical Locality:* None designated (Boucot and Siehl, 1962, list many localities for the genus, most of which represent *Zdimir* Community occurrences as well). It occurs in the Guitang Member of the Beiliu Formation.

*Geographical Distribution:* See Boucot and Siehl (1962). In South China, this community is widely distributed in Sichuan and Guangxi, and southeastern Yunnan (Wang and Yu, 1962; Yu and Kuang, 1986). Wan and others (1978) and Chen Yuanren (written communication, 1980) reported this community from the Eifelian of North Sichuan in the Chayuanzi Member of the Yangmaba Formation.

*Environment:* As discussed by Boucot (1975), this community represents a relatively rough-water, Benthic Assemblage 3 environment. The matrix is commonly calcarenite. The associated stromatoporoids (Yu Chang-min, oral communication, 1980) are commonly massive, the rugose corals are predominantly solitary types, and the favositids are chiefly massive forms. Dr. Yu (oral communication, 1980) observed isolated, articulated specimens in the typical Nandan Facies (basinal) near its boundary with the Beiliu Facies (*Zdimir* Facies) at Luofu, Nandan County, northwestern Guangxi.

*Community Group:* Gypidulinidae.

## *Zygospiraella–Brachyelasma* Community (Fig. 9)

*Name:* This community is named here.

*Composition:* The fauna consists of abundant, articulated brachiopods (*Zygospiraella, Nalivkinia, Kritorhynchia, Eospirifer,* and *Beitaia*) associated with four genera of less abundant rugose corals and a single genus of stromatoporoid (Ge and others, 1979).

*Age:* The *Zygospiraella-Brachyelasma* Community is of Middle Llandoverian age, as outlined by Rong and Yang (1981).

*Typical Locality:* Leijiatun (see "*Beitaia-Eospirifer* Community" for stratigraphic position).

*Geographic Distribution:* At least four localities.

*Environment:* This community occurs in well-bedded limestone and calcareous mudstone, composing an interval up to 8 m thick. Most brachiopods are articulated. The presence of a medium-diversity, articulated brachiopod fauna with both rugose corals and a stromatoporoid is consistent with Benthic Assemblage 3, moderately quiet-water conditions, as is the communities stratigraphic position. *Zygospiraella* is the most abundant brachiopod in this community, which is similar to *Zygospiraella*-rich communities in both Esthonia and the Siberian Platform. ***Community Group:*** *Striispirifer.*

## ALPHABETICAL LIST OF ILLUSTRATED BRACHIOPOD TAXA, WITH PLATE LOCATIONS

*Acrospirifer* sp. Plate XV, Figs. 29-32.

*Acrospirifer fongi* Plate XIX, Figs. 1-4.

*Acrospirifer houershanensis* Plate XIX, Figs. 9-11.

*Aegiria grayi* Plate VI, Figs. 17-18, 23, 33.

*Aegiria shiqianensis* Plate VI, Figs. 19-22, 24.

*Aegiromena convexa* Plate II, Figs. 1-3.

*Aesopomum delicatum* Plate VIII, Figs. 1-3, 14-22, 32.

*Amboglossa transversa* Plate XX, Figs. 12-15.

*Amboglossa waganovae* Plate XX, Figs. 17-20.

*Antirhynchonella* cf. *lingulifera* Plate V, Figs. 8, 11, 12, 15.

*Aphanomena parvicostellata* Plate II, Figs. 8-10.

*Aseptalium kwangsiensis* Plate XI, Figs. 1-9.

*Athyris* sp. Plate XIV, Figs. 18-20.

*Athyris subpentagona* Plate XV, Figs. 9-12.

*Athyrisina simplex* Plate XV, Figs. 17-20.

*Athyrisina squamosaeformis* Plate XVIII, Figs. 6, 8, 10; Plate XIX, Fig. 8.

*Atrypa* sp. Plate XIV, Figs. 9-12.

*Atrypinopsis biconvexa* Plate III, Figs. 11-14.

*Atrypoidea inflata* Plate VIII, Figs. 23-29.

*Atrypoidea lentiformis* Plate VII, Figs. 26-29.

*Atrypopsis rongxiensis* Plate VII, Figs. 10-13.

*Beitaia modica* Plate III, Figs. 29-31.

*Borealis borealis* Plate V, Figs. 1-4.

*Brevilamnulella undata* Plate IV, Figs. 1-5.

*Brevilamnulella* sp. Plate VII, Figs. 1-3.

*Carinatina* cf. *arimaspa* Plate XVII, Figs. 4-5, 9, 14.

*Coolinia* sp. Plate II, Fig. 20.

*Cryptatrypa ovata* Plate VII, Figs. 5-9.

*Cyrtina pingnanensis* Plate XX, Figs. 1-4, 28.

*Dalmanella testudinaria* Plate I, Figs. 1-3.

*Desquamatia hemisphaerica* Plate XVII, Figs. 1-3, 6-8.

*Devonalosia ertangensis* Plate XVI, Figs. 25-27.

*Dicoelostrophia crenata* Plate XII, Figs. 23-24; Plate XIII, Fig. 23.

*Dicoelostrophia punctata* Plate XII, Figs. 18-21.

*Draborthis caelebs* Plate I, Figs. 8-9, 16, 18.

*Dysprosorthis sinensis* Plate I, Figs. 5, 19, 21.

*Eifelatrypa superplana* Plate XVII, Figs. 12, 13.

*Eosophragmophora sinensis* Plate XII, Figs. 1-6.

*Eospiriferina lachrymosa* Plate XVIII, Figs. 1-5, 7, 17.

*Euryspirifer qijianensis* Plate XIX, Figs. 12-15.

*Gypidula longdongshuiensis* Plate XVI, Figs. 16-24.

*Harpidium dorsoplanus,* Plate IV, Figs. 6-9.

*Hindella crassa incipiens* Plate II, Figs. 16-17.

*Hirnantia sagittifera* Plate I, Figs. 11-13, 17.

*Howellella fecunda* Plate XV, Figs. 25-28.

*Howellella shiqianensis* Plate VII, Figs. 30-32.

*Howellella tenuiplicata* Plate XV, Figs. 21-22.

*Howellella tingi* Plate IX, Figs. 19-23.

*Howellella yujiangensis* Plate XIV, Figs. 1-5.

*Huananochonetes ovalis* Plate XII, Figs. 25-26.

*Indospirifer quadriplicatus* Plate XX, Figs. 7-8, 11.

*Kinnella kielanae* Plate I, Figs. 6-7, 14.

*Kwangsia perfecta* Plate XVIII, Figs. 9, 11-16, 18.

*Latonotoechia parasappho* Pl. XIII, Figs. 18-21.

*Leptaenopoma trifidum* Plate II, Figs. 5, 7, 18.

*Leptostrophia elegestica* Plate X, Figs. 20-21.

*Levenea depressa* Plate XII, Figs. 7-11.

*Levenea qianbeiensis* Plate III, Figs. 20-24.

*Linostrophomena convexa* Plate VI, Figs. 34-35.

*Lissatrypa magna* Plate III, Figs. 32-35.

*Longdongshuia subaequata* Plate XVI, Figs. 11-15.

*Luanquella kwangsiensis* Plate XII, Fig. 30; XIII, Figs. 14-15, 24.

*Megaspinochonetes subrectangularis* Plate VI, Figs. 12-16.

*Megastrophia sphaeroidea* Plate XII, Figs. 17, 22, 27.

*Merciella striata* Plate III, Figs. 1-6.

*Meristina barkolensis* Plate X, Figs. 12-14, 19.

*Mirorthis mira* Plate I, Figs. 10, 15, 20.

*Molongia gashaomiaoensis* Plate X, Figs. 1-3, 6.

*Nalivkinia magna* Plate VII, Figs. 14-23.

*Orientospirifer minor* Plate XV, Figs. 1-4.

*Orientospirifer nahkaolingensis* Plate XI, Figs. 25-31.

*"Orientospirifer" wangi* Plate XI, Figs. 20-24, 32.

*Parachonetes nasutus* Plate XIII, Figs. 22, 25-26.

*Paraconchidium shiqianensis* Plate IV, Figs. 16, 19-22; Plate V, Fig. 6

*Parathyrisina* sp. Plate XV, Figs. 5-8.

*Paromalomena polonica* Plate II, Figs. 4, 6.

*Pholidostrophia (Mesopholidostrophia) minor* Plate VI, Figs. 30-32.

*Planatrypa guangxiensis* Plate XVII, Figs. 10-11, 15-18.

*Plectothyrella crassicosta* Plate II, Figs. 11-15, 19.

*Pleurodium tenuiplicatum* Plate IV, Figs. 10, 12-13, 17-18.

*Plicidium sinanensis* Plate III, Figs. 38-41.

*Protathyris xungmiaoensis* Plate VIII, Figs. 8-11.

*Protathyrisina minor* Plate IX, Figs. 27-28, 30-31, 35.

*Protathyrisina plicata* Plate IX, Figs. 29, 32-34.

*Protathyrisina prominula* Plate IX, Figs. 9-11.

*Protathyrisina puta* Plate IX, Figs. 36-40.

*Protathyrisina uniplicata* Plate IX, Figs. 12-15, 24-26.
*Punctatrypa (Undatrypa) bellatula* Plate XIII, Figs. 11-13, 16-17.
*Rensselandia liufengensis* Plate XX, Figs. 22-24.
*Reticulariopsis ertangensis* Plate XV, Figs. 13-16, 23-24.
*Rhipidothyris bicostata* Plate XX, Figs. 9-10, 16, 27.
*Rhipidothyris ovata* Plate XX, Figs. 5-6.
*Rostrospirifer increbescens* Plate XIV, Figs. 13-17.
*Rostrospirifer subtonkinensis* Plate XIX, Figs. 16-18.
*Rostrospirifer tonkinensis* Plate XIV, Figs. 6-8.
*Salopinella minuta* Plate VI, Figs. 1-4.
*Schizophoria communis* Plate XII, Figs. 12-16.
*Schizophoria (Eoschizophoria) hesta* Plate VIII, Figs. 4-7, 12-13.
*Sinochonetes minutisulcatus* Plate XI, Figs. 10-19.
*Spinochonetes notata* Plate VI, Figs. 5-11.
*Spinocyrtia jiuyanensis* Plate XIX, Figs. 5-7.
*Spinulicosta elongata* Plate XVI, Fig. 10.
*Spirigerina sinensis* Plate III, Figs. 25-28, 36-37.
*Spirinella asiatica* Plate IX, Figs. 16-18.
*Spirinella biplicata* Plate VIII, Fig. 31; Plate IX, Figs. 1-4.
*Spirinella sparsa* Plate VIII, Fig. 30; Plate IX, Figs. 5-8.
*Stegerhynchus angaciensis* Plate X, Figs. 4-5.
*Steinhagella guangxiensis* Plate XVI, Figs. 2-4, 28-29.
*Stricklandia transversa* Plate V, Figs. 5, 7, 9-10, 13-14, 16-18.

*Stricklandiella robusta* Plate IV, Figs. 11, 14-15.
*Striispirifer* sp. Plate VII, Figs. 24-25.
*Striispirifer acuminiplicatus* Plate III, Figs. 7-10.
*Striispirifer bellatullus* Plate VII, Figs. 33-36.
*Stropheodonta*? *luzhaiensis* Plate XVI, Fig. 1.
*Tadschikia tecta* Plate IX, Fig. 41
*Tannuspirifer* cf. *pedaschenkoi* Plate X, Figs. 7-8, 15.
*Toxorthis mirabilis* Plate I, Figs. 4, 22.
*Tuvaella barkolensis* Plate X, Figs. 16-18.
*Tuvaella gigantea* Plate X, Figs. 9-11.
*Uncinulus fasciger* Plate XIII, Figs. 1-5.
*Uncinulus mesodeflectus* Plate XIII, Figs. 6-10.
*Uncinulus wuxuanensis* Plate XVI, Figs. 5-9.
*Valdaria lauta* Plate VI, Figs. 25-29.
*Xenostrophia yukiangensis* Plate XII, Figs. 28-29.
*Xinanospirifer flabellum* Plate VII, Figs. 37-39.
*Zygospiraella crassicosta* Plate III, Figs. 17-19.
*Zygospiraella venusta* Plate III, Figs. 15-16.
*Zdimir beiliuensis* Plate XVII, Fig. 19; Plate XX, Figs. 30-31.
*Zdimir pseudobaschkiricus* Plate XX, Figs. 21, 25-26.
*Zdimir strachovi* Plate XVII, Fig. 20; Plate XX, Fig. 29.
*Zdimir triangulicostatus* Plate XVII, Fig. 21.

## Plate I

## (Hirnantian–Late Ashgillian)

All species illustrated herein came from the Kuanyinchiao Beds (Hirnantian, latest Ordovician) in southern China. Consult the following references for additional information about many of the fossils illustrated in the following plates: Hou (1959), Rong and others (1974), Wang and Rong (1979), Yang and Rong (1982).

1-3. *Dalmanella testudinaria* (Dalman)
  ×2; 1, 3. Tongxi, Guizhou; 2. Bijie, Guizhou.

4, 22. *Toxorthis mirabilis* Rong
  4, ×15; 22, ×40, Yichang, Hubei.

5, 19, 21. *Dysprosorthis sinensis* Rong
  5, 19, ×6; 21, ×5, Yichang, Hubei.

6, 7, 14. *Kinnella kielanae* (Temple)
  6, 7, ×8.5; 14, ×8, Yichang, Hubei.

8, 9, 16, 18. *Draborthis caelebs* Marek and Havlicek
  8, 9, ×4; 16, 18, ×2, Yichang, Hubei.

10, 15, 20. *Mirorthis mira* Zeng
  10, ×3; 15, 20, ×4, Yichang, Hubei.

11-13, 17. *Hirnantia sagittifera* (M'Coy)
  ×.5, Yichang, Hubei.

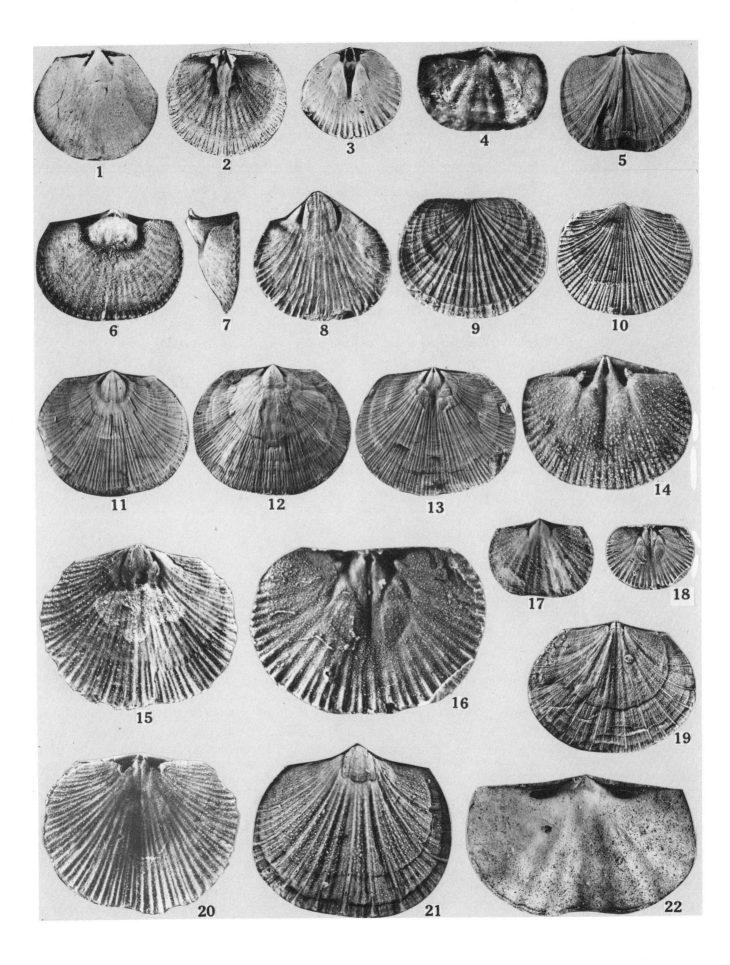

**Plate II**

**(Hirnantian–Late Ashgillian)**

All specimens illustrated come from the Kuanyinchiao Beds in southern China.

1-3. *Aegiromena convexa* Chang
×4, Yichang, Hubei.

4, 6. *Paromalomena polonica* (Temple)
×4, Yichang, Hubei.

5, 7, 18. *Leptaenopoma trifidum* Marek and Havlicek
5, 7, ×2; 18, ×5, Yichang, Hubei.

8-10. *Aphanomena parvicostellata* Rong
8, 9, ×1.5; 10, ×20, Yichang, Hubei.

11-15, 19. *Plectothyrella crassicosta* (Dalman)
×2, 11-13, 19, Bijie, Guizhou; 14, ×2, Tongzi, Guizhou; 15, ×2, Yichang, Hubei.

16, 17. *Hindella crassa incipiens* (Williams)
16, ×2, Yichang, Hubei; 17, ×2, Changning, Sichuan

20. *Coolinia* sp.
×10, Changning, Sichuan.

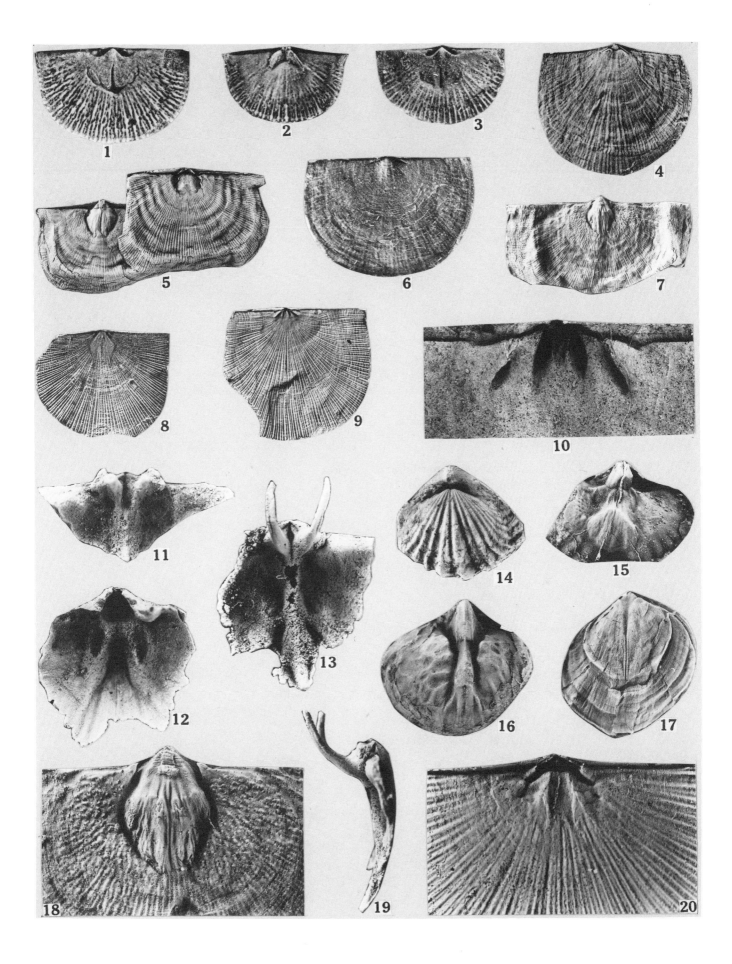

## Plate III

### (Llandoverian)

1-6. *Merciella striata* Rong, Xu and Yang
×2, Lojoping Formation (Late Llandoverian), Yichang, Hubei.

7-10. *Striispirifer acuminiplicatus* Rong and Yang
×2, Lower Xiangshuyuan Formation (Middle Llandoverian), Shiqian, Guizhou.

11-14. *Atrypinopsis biconvexa* Rong and Yang
×3, Xiangshuyuan Formation (Middle-Late Llandoverian), Shiqian, Guizhou.

15, 16. *Zygospiraella venusta* Rong and Yang
×2, Lower Xiangshuyuan Formation (Middle Llandoverian), Yinjiang, Guizhou.

17-19. *Zygospiraella crassicosta* Rong and Yang
×2, Lojoping Formation (Late Llandoverian), Yichang, Hubei.

20-24. *Levenea qianbeiensis* (Rong and Yang)
×2, Lojoping Formation (Late Llandoverian), Yichang, Hubei.

25-28, 36, 37. *Spirigerina sinensis* (Wang)
25-28, ×1.5; 36, 37, ×8, Lower Xiangshuyuan Formation (Middle Llandoverian), Shiqian, Guizhou.

29-31. *Beitaia modica* Rong, Xu, Wu and Yang
×1.5, Lower Xiangshuyuan Formation (Middle Llandoverian), Shiqian, Guizhou.

32-35. *Lissatrypa magna* (Grabau)
32, 33, ×1.5; 34, 35, ×2, Lojoping Formation (Late Llandoverian), Yichang, Hubei.

38-41. *Plicidium sinanensis* (Rong and Yang)
×2, Xiangshuyuan Formation (Middle-Late Llandoverian), Sinan, Guizhou.

## Plate IV

### (Llandoverian)

1-5. *Brevilamnulella undata* (Sowerby)
    ×2, Upper Xiangshuyuan Formation (Late Llandoverian), Yinjiang, Guizhou.

6-9. *Harpidium dorsoplanus* (Wang)
    ×1, Lojoping Formation (Late Llandoverian), Yichang, Hubei.

10, 12, 13, 17, 18. *Pleurodium tenuiplicatum* (Grabau)
    ×1. Lojoping Formation (Late Llandoverian), Yichang, Hubei.

11, 14, 15. *Stricklandiella robusta* Rong and Yang
    ×1.5, Lojoping Formation (Late Llandoverian), Yichang, Hubei.

16, 19-22. *Paraconchidium shiqianensis* Rong, Xu, Fang and Yang
    16, 19-21, ×1; 22, ×2, Upper Xiangshuyuan Formation (Late Llandoverian), Shiqian,
    Guizhou.

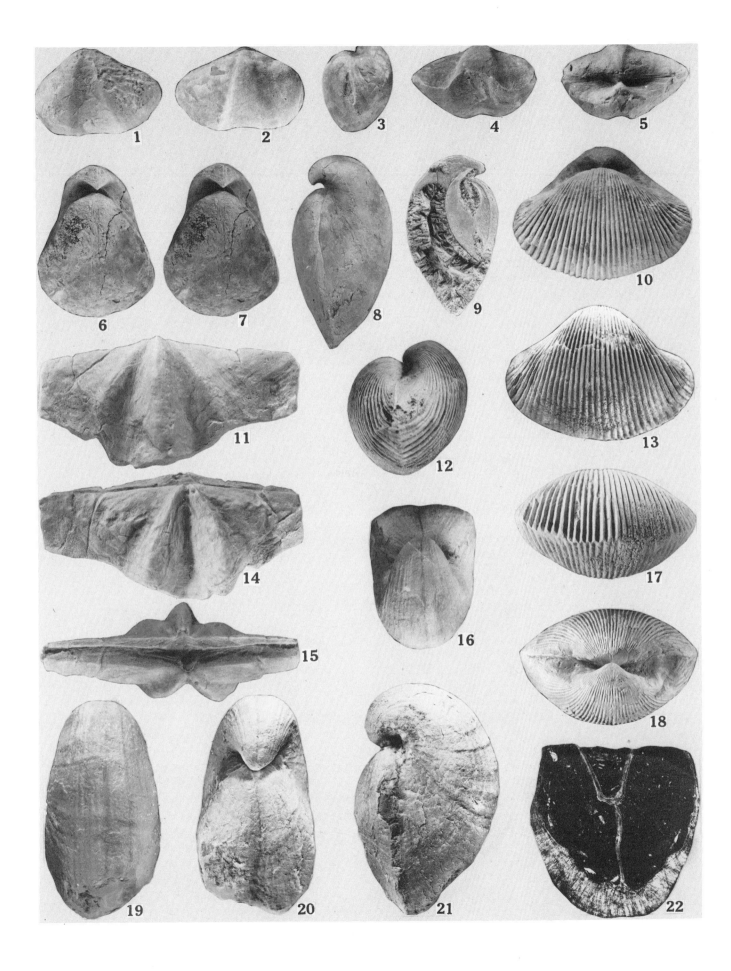

**Plate V**

**(Llandoverian)**

1-4. *Borealis borealis* (Eichwald)
  1, 2, ×1; 3, ×2; 4, ×3, Lower Xiangshuyuan Formation (Middle Llandoverian), Sinan Guizhou.

5, 7, 9, 10, 13, 14, 16-18. *Stricklandia transversa* Grabau
  5, ×6, Upper Xiangshuyuan Formation (Late Llandoverian), Yinjiang, Guizhou; 7, 10, 14, 16, 18, ×1, Upper Xiangshuyuan Formation (Late Llandoverian), Yinjiang, Guizhou; 9, 17, ×6, Leijiatun Formation (Late Llandoverian), Wuchuan, Guizhou; 13, ×6, Lojoping Formation (Late Llandoverian), Yichang, Hubei.

6. *Paraconchidium shiqianensis* Rong, Xu, Fang and Yang
  ×4, Upper Xiangshuyuan Formation (Late Llandoverian), Shiqian, Guizhou.

8, 11, 12, 15. *Antirhynchonella* cf. *lingulifera* (Sowerby)
  ×1.5, "Shihniulan Formation" (Late Llandoverian), Yanhe, Guizhou.

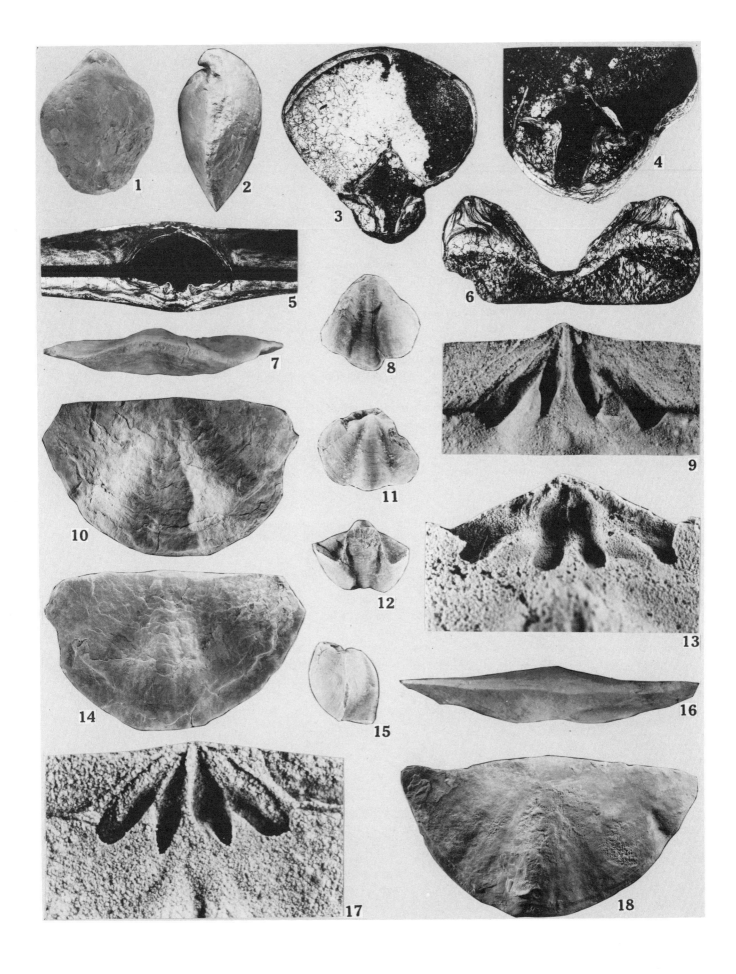

## Plate VI

### (Latest Llandoverian–Early Wenlockian)

All specimens illustrated here came from the Upper Xiushan Formation (Latest Llandoverian–Early Wenlockian).

1-4. *Salopinella minuta* (Rong and Yang)
    1, ×8; others ×10, 1-3, Rongxi, Xiushan, Sichuan; 4, Siyangqiao, Badong, SW Hubei.

5-11. *Spinochonetes notata* Liu and Xu
    5, 6, 8, 11, ×6; 7, 9, ×10; 10, ×8, Rongxi, Xiushan, SE Sichuan.

12-16. *Megaspinochonetes subrectangularis* Yang and Rong
    15, ×4; others ×3, Leijiatun, Shiqian, NE Guizhou.

17, 18, 23, 33. *Aegiria grayi* (Davidson)
    17, ×8, Rongxi, Xiushan, SE Sichuan; 18, 23, 33, ×8, ×10, ×8, Gaoluo, Xuanen, SW Hubei.

19-22, 24. *Aegiria shiqianensis* Yang and Rong
    19, 20, 22, ×6, Rongxi, Xiushan, SE Sichuan; 21, ×10, Gaoluo, Xuanen, SW Hubei; 24, ×8, Leijiatun, Shiqian, NE Guizhou.

25-29. *Valdaria lauta* (Rong and Yang)
    25, 26, ×1.5, ×2, Gaoluo, Xuanen, SW Hubei; 27, ×2, Juntianba, Shiqian, NE Guizhou; 28, 29, ×1.5, ×6, Leijiatun, Shiqian, NE Guizhou.

30-32. *Pholidostrophia (Mesopholidostrophia) minor* (Xu and Rong)
    30, 31, ×2, Gaoluo, Xuanen, SW Hubei; 32, ×8, Leijiatun, Shiqian, NE Guizhou.

34, 35. *Linostrophomena convexa* (Xu and Rong)
    ×4, ×8, Rongxi, Xiushan, SE Sichuan.

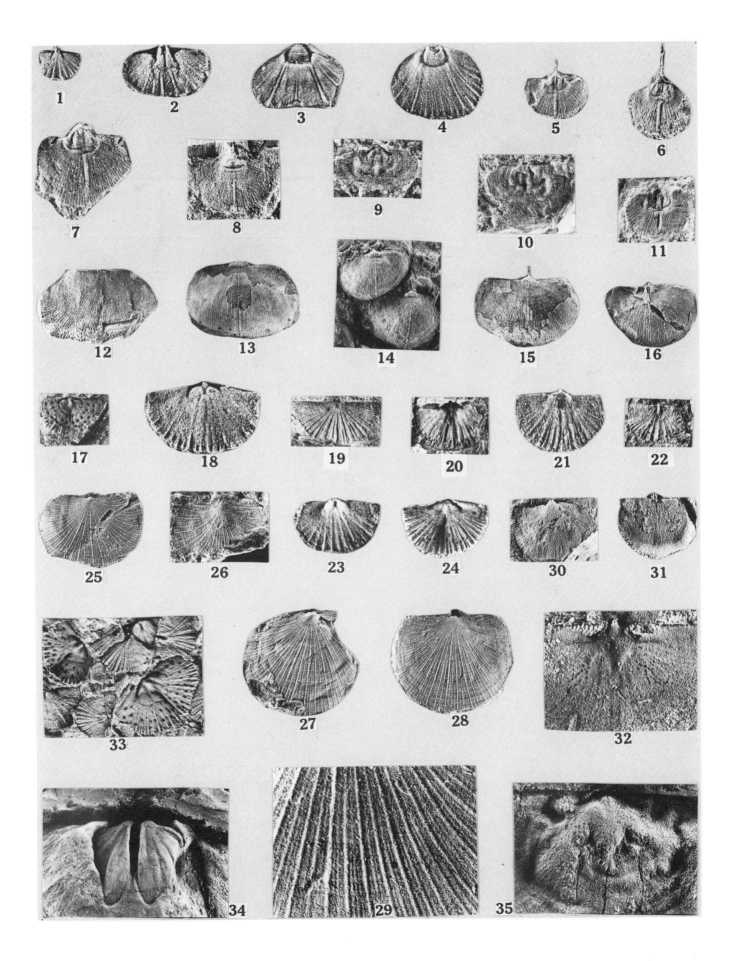

**Plate VII**

**(Latest Llandoverian–Early Wenlockian)**

All specimens illustrated here came from the Upper Xiushan Formation (Latest Llandoverian–Early Wenlockian).

1-3. *Brevilamnulella* sp.
    ×2, Juntianba, Shiqian, NE Guizhou.

5-9. *Cryptatrypa ovata* Yang and Rong
    5, 6, ×2; 7-9, ×6, Rongxi, Xiushan, SE Sichuan.

10-13. *Atrypopsis rongxiensis* (Wan)
    ×1, Rongxi, Xiushan, Sichuan.

14-23. *Nalivkinia magna* Yang and Rong
    14-16, ×1.5; 17-21, ×1; 22, 23, ×1.5, Leijiatun, Shiqian, NE Guizhou.

24, 25. *Striispirifer* sp.
    ×3, Rongxi, Xiushan, SE Sichuan.

26-29. *Atrypoidea lentiformis* (Wang)
    ×2, Shuanghe, Changning, SW Sichuan.

30-32. *Howellella shiqianensis* (Rong and Yang)
    ×1.5, Heshui, Yinjiang, NE Guizhou.

33-36. *Striispirifer bellatullus* (Rong and Yang)
    ×3, Rongxi, Xiushan, SE Sichuan.

37-39. *Xinanospirifer flabellum* Rong, Xu, Sun and Yang
    37, ×2; 38, ×8, Tongguyuan, Shiqian, NE Guizhou.

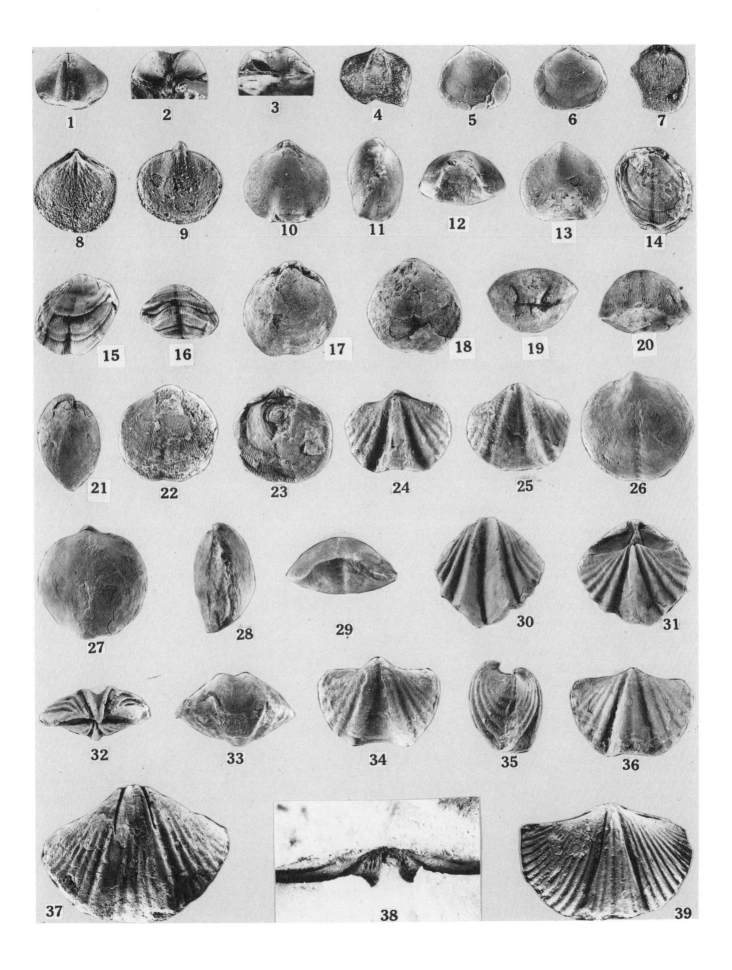

## Plate VIII

### (Late Ludlovian)

All specimens illustrated here came from the Miaokao Formation (Late Ludlovian), Qujing, Eastern Yunnan.

1-3, 14-22, 32. *Aesopomum delicatum* (Rong, Xu and Yang)
    1-3, 14-19, ×1.5; 20-22, ×2; 32, ×8.

4-7, 12, 13. *Schizophoria (Eoschizophoria) hesta* Rong and Yang
    ×2.

8-11. *Protathyris xungmiaoensis* Chu
    ×1.5

23-29. *Atrypoidea inflata* (Fang)
    23, 24, 28, 29, ×1; 25-27, ×2.

30. *Spirinella sparsa* Rong and Yang
    ×8.

31. *Spirinella biplicata* (Chu)
    ×8.

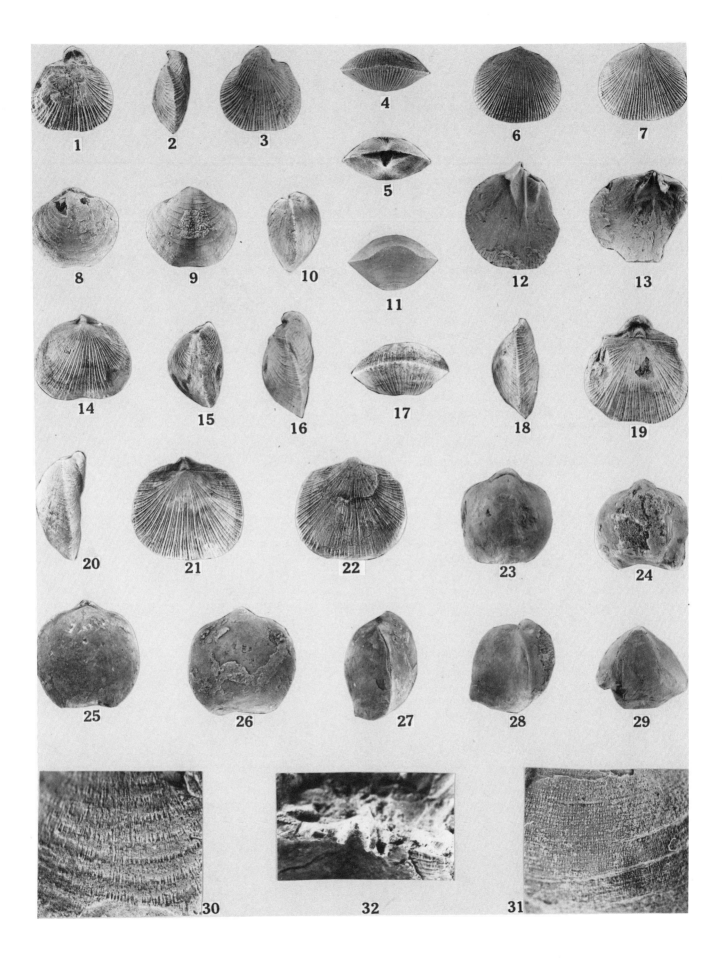

**Plate IX**

**(Late Ludlovian)**

All specimens illustrated here came from the Miaokao Formation (Late Ludlovian), Qujing, eastern Yunnan.

1-4. *Spirinella biplicata* (Chu)
    ×1.

5-8. *Spirinella sparsa* Rong and Yang
    ×2.

9-11. *Protathyrisina prominula* Rong and Yang
    ×2.

12-15, 24-26. *Protathyrisina uniplicata* (Grabau)
    ×2.

16-18. *Spirinella asiatica* Rong and Yang
    ×1.

19-23. *Howellella tingi* (Grabau)
    ×2.

27, 28, 30, 31, 35. *Protathyrisina minor* (Hayasaka)
    ×2.

29, 32-34. *Protathyrisina plicata* (Mansuy)
    ×2.

36-40. *Protathyrisina puta* (Rong and Yang)
    38, ×4; others ×1.5.

41. *Tadschikia tecta* Rong and Yang
    ×1.5.

## Plate X

### (Ludlovian–Pridolian)

1-3, 6. *Molongia gashaomiaoensis* Su
  ×3, Xibiehe Formation (Pridolian), Gashaomiao, Damao Qi, central Inner Mongolia.

4, 5. *Stegerhynchus angaciensis* Chernyshev)
  ×3, Gashaomiao Formation (Pridolian), Gashaomiao, Damao Qi, central Inner Mongolia.

7, 8, 15. *Tannuspirifer* cf. *T. pedaschenkoi* (Chernyshev)
  ×2, Hongliuxia Formation (Ludlovian), Barkol, NE Xinjiang.

9-11. *Tuvaella gigantea* (Chernyshev)
  ×2, Hongliuxia Formation (Ludlovian), Barkol, NE Xinjiang.

12-14, 19. *Meristina barkolensis* Zhang
  ×2, Hongliuxia Formation (Ludlovian), Barkol, NE Xinjiang.

16-18. *Tuvaella barkolensis* Zhang, Rong and Di
  ×2, Hongliuxia Formation (Ludlovian), Barkol, NE Xinjiang.

20, 21. *Leptostrophia elegestica* Chernyshev
  ×1.5, Hongliuxia Formation (Ludlovian), Barkol, NE Xinjiang.

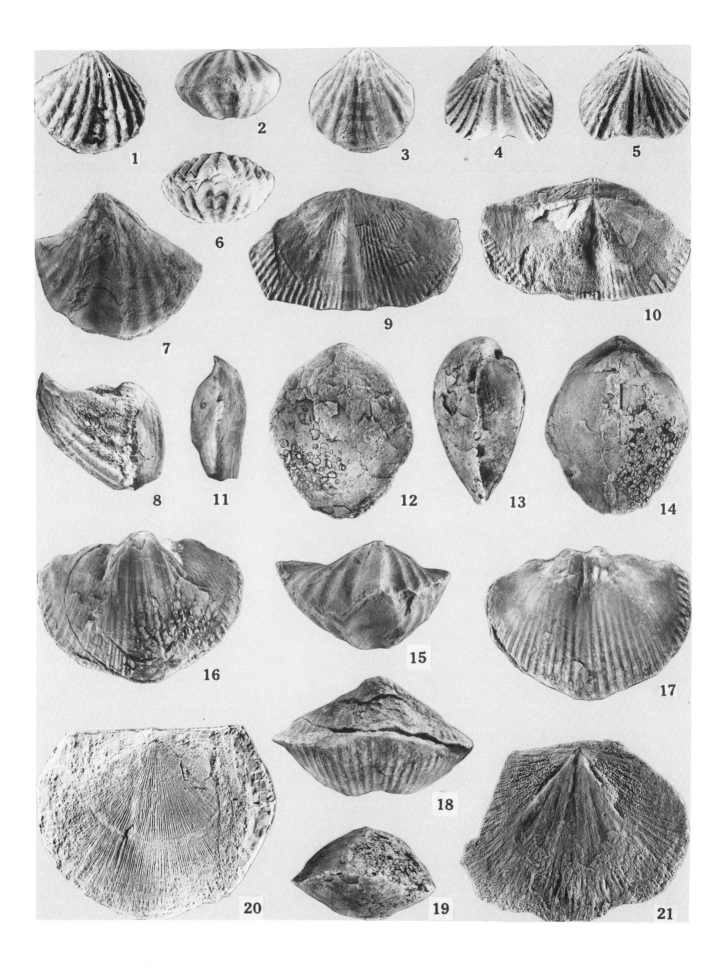

## Plate XI

### (Late Siegenian)

All specimens illustrated here came from the Nahkaoling Formation (Late Siegenian) of Liujing, Henxian, southern Guangxi.

1-9. *Aseptalium kwangsiensis* (Wang)
×2.

10-19. *Sinochonetes minutisulcatus* Wang, Boucot and Rong
×3.

20-24, 32. *"Orientospirifer" wangi* (Hou)
×2.

25-31. *Orientospirifer nahkaolingensis* (Wang)
×2.

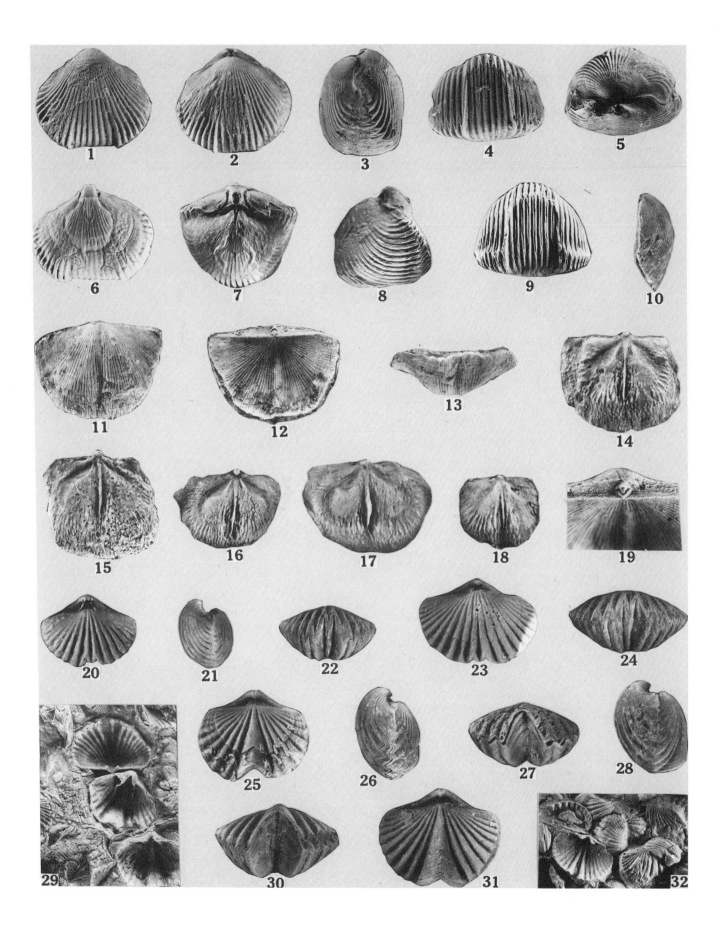

*Wang Yu and Others*

**Plate XII**

**(Early Emsian)**

All specimens illustrated here came from the Yukiang Formation (Early Emsian) of Liujing, Hengxian, central Guangxi.

1-6. *Eosophragmophora sinensis* Wang
1-3, ×5; 6, ×2.5; 4, ×3.

7-11. *Levenea depressa* Wang
×2.5.

12-16. *Schizophoria communis* (Yin)
×2.

17, 22, 27. *Megastrophia sphaeroidea* (Hou and Xian)
×1.

18-21. *Dicoelostrophia punctata* Wang
×1.5.

23-24. *Dicoelostrophia crenata* Wang
×1.5.

25-26. *Huananochonetes ovalis* (Hou and Xian)
×3.

28-29. *Xenostrophia yukiangensis* (Wang)
×2.

30. *Luanquella kwangsiensis* (Wang)
×1.5.

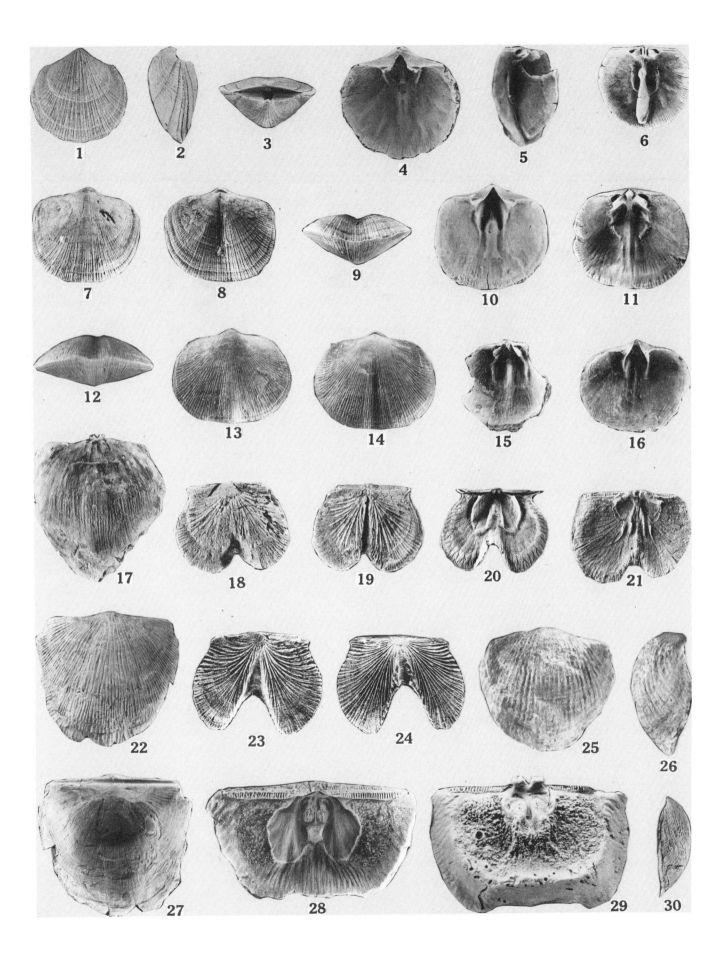

**Plate XIII**

**(Early Emsian)**

All specimens illustrated here came from the Yukiang Formation (Early Emsian) of Liujing, Hengxian, central Guangxi.

1-5. *Uncinulus fasciger* Hou and Xian
    ×2.

6-10. *Uncinulus mesodeflectus* Wang
    ×2.

11-13, 16-17. *Punctatrypa (Undatrypa) bellatula* Wang, Copper and Rong
    ×3.

14, 15, 24. *Luanquella kwangsiensis* (Wang)
    14, 15, ×1.5; 24, ×2.

18-21. *Latonotoechia parasappho* (Wang)
    ×2.

22, 25, 26. *Parachonetes nasutus* Wang
    ×1.5.

×23. *Dicoelostrophia crenata* Wang
    ×1.5.

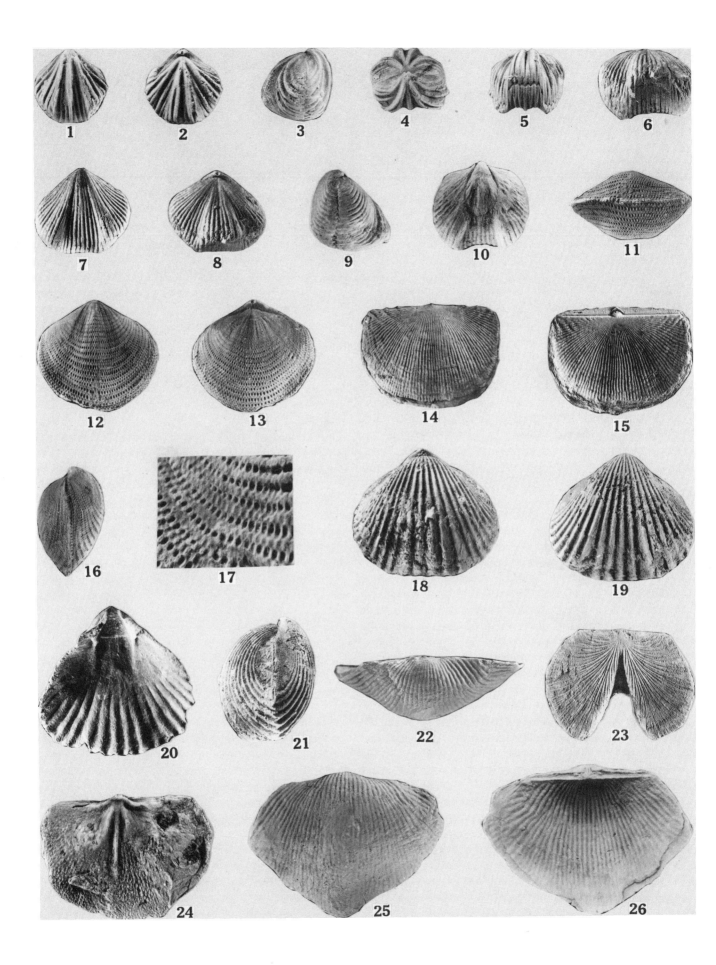

**Plate XIV**

**(Early Emsian)**

All specimens illustrated here came from the Yukiang Formation (Early Emsian) of Liujing, Hengxian, central Guangxi.

1-5. *Howellella yujiangensis* Hou and Xian
    1-4, ×2.5; 5, ×10.

6-8. *Rostrospirifer tonkinensis* (Mansuy)
    ×1.

9-12. *Atrypa* sp.
    ×1.5.

13-17. *Rostrospirifer increbescens* (Wang)
    ×1.

18-20. *Athyris* sp.
    ×1.5.

**Plate XV**

**(Late Early Emsian)**

All specimens illustrated, except for Figs. 29-32, came from the Ertang Formation (Late Early Emsian), Ertang, Wuxuan, Guangxi.

1-4. *Orientospirifer minor* Chen
      ×2.5.

5-8. *Parathyrisina* sp.
      ×2.5.

9-12. *Athyris subpentagona* Chen
      ×1.5.

13-16, 23, 24. *Reticulariopsis ertangensis* Chen
      ×1.5.

17-20. *Athyrisina simplex* Chen
      ×2.5.

21, 22. *Howellella tenuiplicata* Chen
      ×1.5.

25-28. *Howellella fecunda* Chen
      ×1.5.

29-32. *Acrospirifer* sp.
      ×1.5, "Ertang Formation," Dale, Xiangzhou, Guangxi.

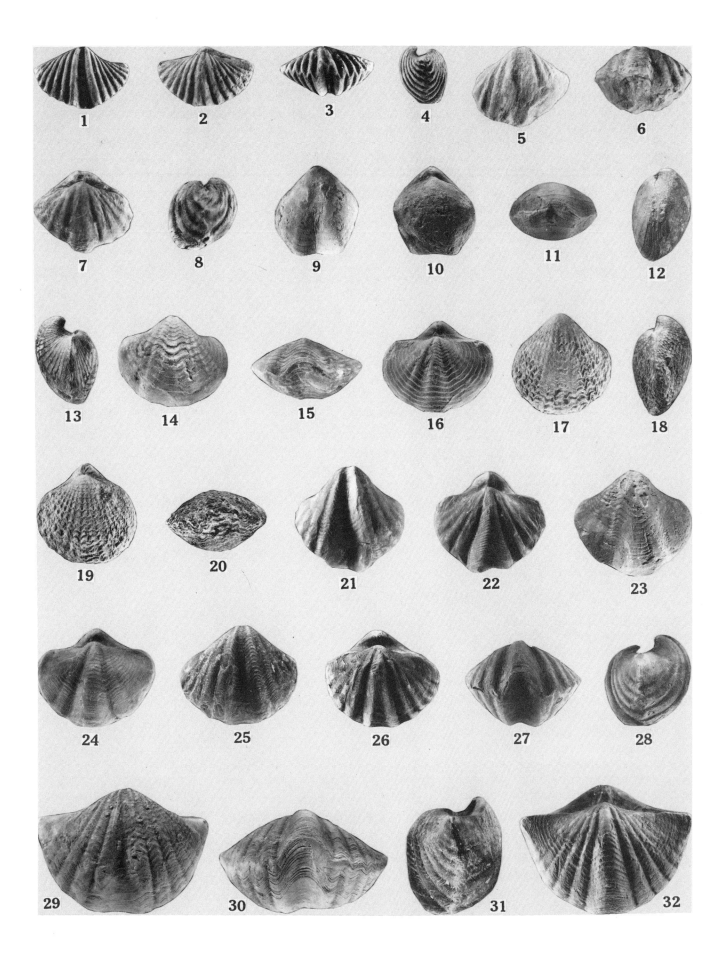

## Plate XVI

### (Late Emsian–Eifelian)

1. *Stropheodonta? luzhaiensis* Wang and Zhu
   ×1.5, Sipai Formation (Late Emsian), Sipai, Luzhai, Guangxi.

2-4, 28, 29. *Steinhagella guangxiensis* Wang and Zhu
   2-4, ×1.5; 28, 29, ×2, Baima Shale (Eifelian), Pingnan, Guangxi.

5-9. *Uncinulus wuxuanensis* Wang and Zhu
   ×3, Laohuling Member (Eifelian) of the Sipai Formation, Ertang, Wuxuan, Guangxi.

10. *Spinulicosta elongata* Wang and Zhu
    ×2, Longdongshui Member of the Houershan Formation (Eifelian), Dushan, Guizhou.

11-15. *Longdongshuia subaequata* Hou and Xian
    11-14, ×2; 15, ×1, Longdongshui Member of the Houershan Formation (Eifelian), Dushan, Guizhou.

16-24. *Gypidula longdongshuiensis* (Wang, Liu, Wu, and Zhong)
    ×2, Longdongshui Member (Eifelian) of the Houershan Formation, Dushan, Guizhou.

25-27. *Devonalosia ertangensis* Wang and Zhu
    ×3, Laohuling Member (Eifelian) of the "Sipai Formation," Ertang, Wuxuan, Guangxi.

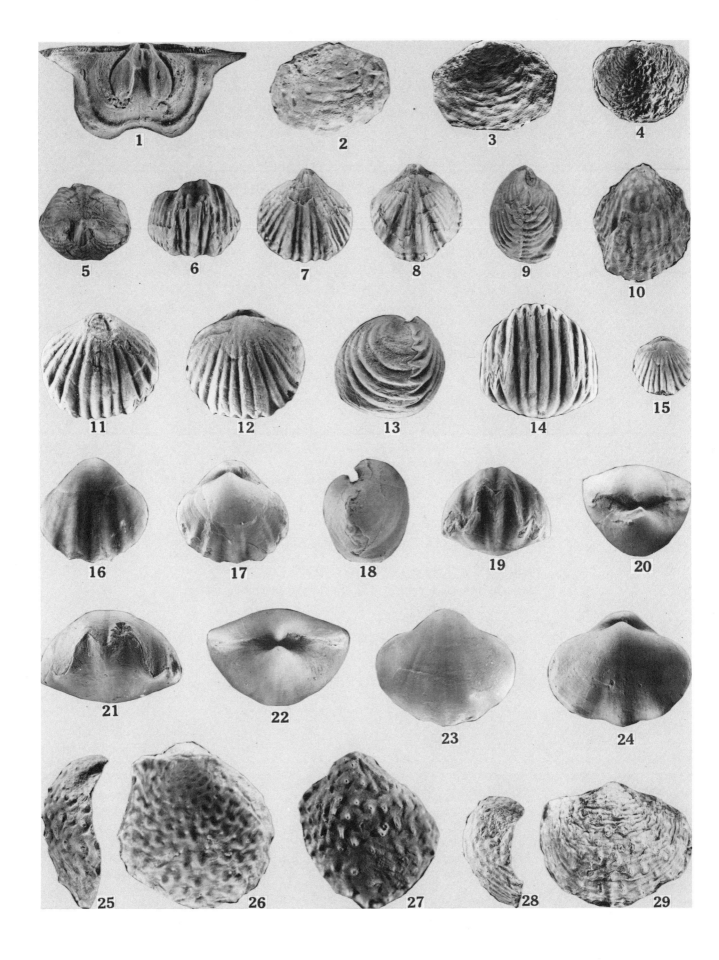

## Plate XVII

### (Late Emsian–Eifelian)

1-3, 6-8. *Desquamatia hemisphaerica* Wang and Zhu
    ×1, Longdongshui Member (Eifelian) of the Houershan Formation, Dushan, southeastern Guizhou.

4-5, 9, 14. *Carinatina* cf. *arimaspa* (Eichwald)
    ×2, Guitang Member (Late Emsian) of the Beiliu Formation, Beiliu, Guangxi.

10, 11, 15-18. *Planatrypa guangxiensis* Wang and Zhu
    10, 11, 16-18, ×1.5; 15, ×4, Sipai Formation (Late Emsian–Early Eifelian), Luzhai, Guangxi.

12, 13. *Eifelatrypa superplana* Wang and Zhu
    ×1.5, Guitang Member (Late Emsian) of the Beiliu Formation, Beiliu, Guangxi.

19. *Zdimir beiliuensis* Wang and Zhu
    ×1, Guitang Member (Late Emsian) of the Beiliu Formation, Beiliu, Guangxi.

20. *Zdimir strachovi* (Andronov)
    ×1, Guitang Member (Late Emsian) of the Beiliu Formation, Beiliu, Guangxi.

21. *Zdimir triangulicostatus* (Andronov)
    ×1, Guitang Member (Late Emsian) of the Beiliu Formation, Beiliu, Guangxi.

## Plate XVIII

### (Eifelian)

1-5, 7, 17. *Eospiriferina lachrymosa* Grabau
    1-5, ×2; 7, ×4; 17, ×5, Sipai Member (Eifelian) of the Sipai Formation, Qijian, Xiangzhou, Guangxi.

6, 8, 10. *Athyrisinia squamosaeformis* Wang, Liu, Wu, and Zhong
    ×2.5, Longdongshui Member (Eifelian) of the Houershan Formation, Dushan, Guizhou.

9, 11-16, 18. *Kwangsia perfecta* Wang and Zhu
    9, 11, 12, ×2; 13-16, 18, ×1.5, Longdongshui Member (Eifelian) of the Houershan Formation, Dushan, Guizhou.

**Plate XIX**

**(Late Emsian–Eifelian)**

1-4. *Acrospirifer fongi* (Grabau)
 ×1, Longdongshui Member (Eifelian) of the Houershan Formation, Dushan, Guizhou.

5-7. *Spinocyrtia jiuyanensis* Wang and Zhu
 ×1, "Sipai Formation," Ertang, Wuxuan, Guangxi.

8. *Athyrisina squamosaeformis* Wang, Liu, Wu, and Zhong
 ×2.5, Longdongshui Member (Eifelian) of the Houershan Formation, Dushan, Guizhou.

9-11. *Acrospirifer houershanensis* Hou and Xian
 ×1, Longdongshui Member (Eifelian) of the Houershan Formation, Dushan, Guizhou.

12-15. *Euryspirifer qijianensis* Wang, Liu, Wu, and Zhong
 ×1, Luma Member (Late Emsian) of the "Sipai Formation," Qijian, Xiangzhou, Guangxi.

16-18. *Rostrospirifer subtonkinensis* Wang and Wu
 ×1, Luma and Sipai Members (Late Emsian–Eifelian) of the Sipai Formation, Luzhai, Guangxi.

## Plate XX

### (Late Emsian–Eifelian)

1-4, 28. *Cyrtina pingnanensis* Wang and Zhu
    ×2.5, Maima Member (Eifelian), Pinan, Guangxi.

5, 6. *Rhipidothyris ovata* Wang and Zhu
    ×5, Sipai Member (Eifelian) of the "Sipai Formation," Qijian, Xiangzhou, Guangxi.

7, 8, 11. *Indospirifer quadriplicatus* (Chang)
    ×1.5, Luma Member (Late Emsian) of the "Sipai Formation," Qijian, Xiangzhou, Guangxi.

9, 10, 16, 27. *Rhipidothyris bicostata* Wang and Zhu
    ×2.5, Baima Shale (Eifelian), Pingnan, Guangxi.

12-15. *Amboglossa transversa* (Wang, Liu, Wu, and Zhong)
    12-14, ×2.5; 15, ×6, Longdongshui Member (Eifelian) of the Houershan Formation, Dushan, Guizhou.

17-20. *Amboglossa waganovae* (Breivel)
    ×1.5, Laohuling Member (Eifelian) of the "Sipai Formation," Ertang, Wuxuan, Guangxi.

21, 25, 26. *Zdimir pseudobaschkiricus* (Chernyshev)
    ×1, Guitang Member (Late Emsian) of the Beiliu Formation, Beiliu, Guangxi.

22-24. *Rensselandia liufengensis* Wang and Zhu
    ×1, Laohuling Member (Eifelian) of the "Sipai Formation," Ertang, Wuxuan, Guangxi.

29. *Zdimir strachovi* (Andronov)
    ×1, Guitang Member (Eifelian) of the Beiliu Formation, Beiliu, Guangxi.

30, 31. *Zdimir beiliuensis* Wang and Zhu
    ×1, Guitang Member (Eifelian) of the Beiliu Formation, Beiliu, Guangxi.

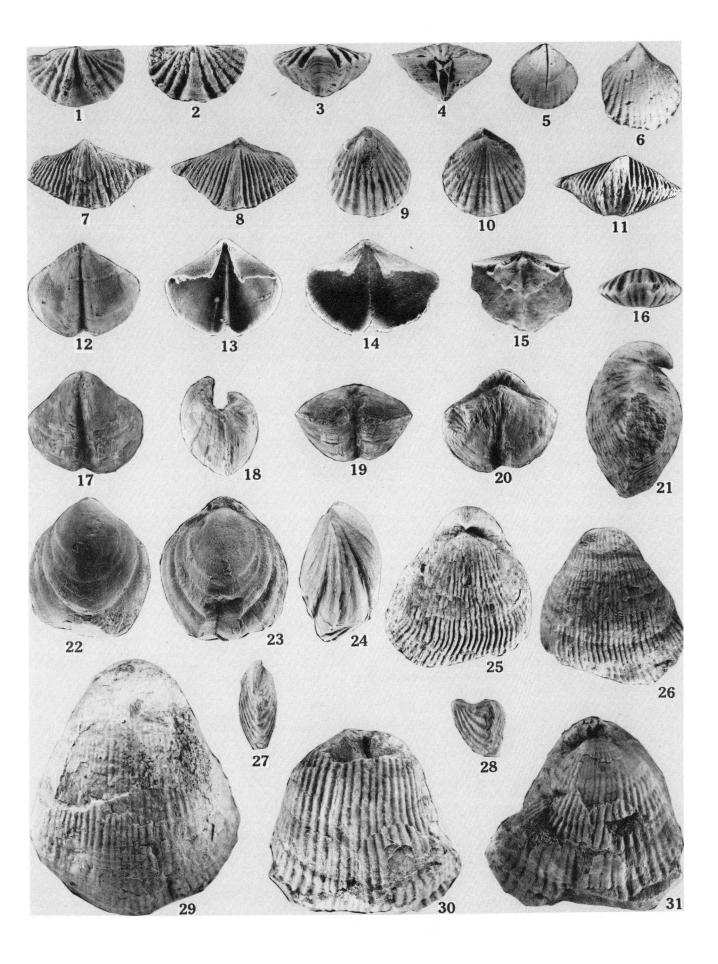

# REFERENCES CITED

Ager, D. V., 1963, Paleoecology: McGraw-Hill, 371 p.

Anderson, M. M., Boucot, A. J., and Johnson, J. G., 1969, Eifelian Brachiopods from Padaukpin, Northern Shan States, Burma: Bulletin of the British Museum (Natural History), Geology, v. 18, no. 4, p. 107–163.

Berry, W.B.N., and Boucot, A. J., 1972, Silurian Graptolite depth zonation: Proceedings of the 24th International Geological Congress, Montreal, Section 7, Paleontology, p. 59–65.

Blieck, A., 1982, Les grandes lignes de la biogéographie des Hetrostraces du Silurien Supérieur; Devonien Inférieur dans le domain Nord-Atlantique: Palaeogeography, Palaeoclimatology, Palaeoecology, v. 38, p. 283–316.

Boucot, A. J., 1975, Evolution and Extinction Rate Controls: Elsevier, 427 p.

——, 1978, Community Evolution and Rates of Cladogenesis: Evolutionary Biology, v. 11, p. 545–655.

——, 1981, Principles of Benthic Marine Paleoecology: Academic Press, 470 p.

——, 1982, Ecostratigraphic framework for the Lower Devonian of the North American Appohimchi Subprovince: Neues Jahrbuch für Geologie und Paläontologie Abhandlungen, v. 163, p. 81–121.

—— 1983, Does evolution take place in an ecologic vacuum?: Journal of Paleontology, v. 57, p. 1–30.

——, 1984, Ecostratigraphy and Autecology in the Silurian: Palaeontological Association Special Papers in Palaeontology, v. 32, p. 7–16.

——, 1985, Late Silurian-early Devonian biogeography, provincialism, evolution, and extinction: Philosophical Transactions of the Royal Society of London, B, v. 309, p. 323–339.

Boucot, A. J., and Johnson, J. G., 1979, Pentamerinae (Silurian brachiopoda): Palaeontographica, Abt. A, v. 163, p. 87–129.

Boucot, A. J., and Siehl, A., 1962, *Zdimir* Barrande (Brachiopoda) redefined: Wiesbaden, Notizbl. Hessische Landesamt für Bodenforschung, v. 90, p. 117–131.

Boucot, A. J., Johnson, J. G., and Struve, W., 1966, *Stringocephalus:* Ontogeny and distribution: Journal of Paleontology, v. 40, p. 1349–1364.

Boucot, A. J., Massa, D., and Perry, D. G., 1983, Stratigraphy, biogeography and taxonomy of some Lower and Middle Devonian brachiopod-bearing beds of Libya and northern Niger: Palaeontographica, Abt. A, Bd. 180, 91–125.

Bretsky, P. W., 1970, Upper Ordovician ecology of the Central Appalachians: Peabody Museum of Natural History, Yale University, Bulletin, v. 34, 150 p.

Chen Yuan-ren, 1979, The fossil brachiopods from the Tudiling member (Bailiuping Formation) of Early Devonian in the Longmenshan Area, Northwestern Sichuan, and their stratigraphic significance: Geologic College of Chengdu (I.R.) Sichuan longmenshanqu zaonipenshi tudilingduan de wanzu dongwu huashi jiqi diceng yiyi, Chengdu dizhi Xueyuan, 1978, p. 7).

Clarke, J. M., 1922, 17th Annual Report of the Director of the New York State Museum and Science Department: Bulletin, v. 239–240, p. 24–25.

Copper, P., 1977a, *Zygospira* and some related Ordovician and Silurian atrypoid brachiopods: Paleontology, v. 20, p. 295–335.

——, 1977b, The late Silurian brachiopod genus *Atrypoidea:* Geologiska Föreningens i Stockholm Förhandlingar, v. 99, p. 10–26.

Feldman, H. R., 1980, Level-bottom brachiopod communities in the Middle Devonian of New York: Lethaia, v. 13, p. 27–46.

Gao Lian-da, 1978, Spores and acritarchs of the Early Devonian Nahkaoling Stage of Liujing, Guangxi, *in* The Symposium on Devonian of South China, edited by Institute of Geological and Mineralogical Research: Chinese Academy of Geological Science, p. 346–358 (Guangxi Liujing Zaonipenshi Nagaoling Jie Baozi he yiyuanlei, Huanan Nipenxi huiy; lunwenji, 346-358 ye. Dizhi Chubanshe).

Ge Zhi-zhou, Rong Jia-yu, Yang Xue-chang, Liu Geng-wu, Ni Yu-nan, Dong De-yuan, and Wu Hong-ji, 1979, The Silurian system in southwest China, *in* The Carbonate Biostratigraphy of Southwest China: Science Press, p. 155–220 (Xinan diqu de Zhiliu xi, Xinan diqu tansuanyan shengwu diceng, 155-220 ye, Kexu Chubanshe).

Hamada, T., 1971, Early Devonian Brachiopods from the Lesser Khingan District of Northeast China: Palaeontological Society of Japan Special Paper 15, 98 p.

Hou Hong-fei, 1959, Spiriferoids of Lower Devonian and Eifelian of S. Guangxi: Acta Palaeontologica Sinica, v. 7, no. 6, p. 450–475 (Guangxi nanbu xianipentong he aifeierjie shiyan huashi, Gushengwu xeubao, 7 Juan, 6 qi, 450-475 ye).

——, 1963, The new genera and species of brachiopods of Middle Devonian: Acta Palaeontologica Sinica, v. 11, no. 3, p. 412–428 (with Russian abstract) (Zhong Nipenshi wanzulei de Xin Shuzhong, Gushengwu Xuebao, 11 Juan, 3 qi, 412-428 ye).

Hou Hong-fei and Xian Si-yuan, 1975, The Brachiopods of Lower and Middle Devonian in Guangxi and Guizhou: Peking, Geological Publishing House, Editorial Committee of Professional Paper of Stratigraphy and Palaeontology, Chinese Academy of Geological Sciences, Professional Papers of Stratigraphy and Palaeontology, no. 1, p. 1–85 (Guangxi, Guizhou Xia Zhong Nipen Tong Wanzulei huashi. Dizhi Kexue Yanjiuyuan, Diceng Gushengwu Lunwenji, 1 ji, 1-85 ye).

Hou Jing-peng, 1978, Chitinozoans of the Devonian Nahkaoling Formation of Liujing, Hengxian, Guangxi, *in* The Symposium on Devonian of South China, edited by Institute of Geological and Mineralogical Research: Chinese Academy of Geological Sicence, p. 359–396 (Guangxi Hengxian Liujing Nipenxi Nagaoling zu Jidingchong, Huanan Nipenxi Lunwen ji, 359-396 ye).

Johnson, J. G., 1970, Early Middle Devonian brachiopods from central Nevada: Journal of Paleontology, v. 44, p. 252–264.

——, 1978, Devonian, Givetian age brachiopods and biostratigraphy, central Nevada: Geologica et Palaeontologica, v. 12, p. 117–150.

Jones, B., 1977, Variation in the Upper Silurian Brachiopod *Atrypella phoca* (Salter) from Somerset and Prince of Wales Islands, Arctic Canada: Journal of Paleontology, v. 51, p. 459–479.

——, 1980, *Tannuspirifer dixoni;* A new species of Spinocyrtiidae from the Read Bay Formation of Somerset Island, Arctic Canada: Journal of Paleontology, v. 54, p. 745–751.

Jones, B., and Rong Jia-yu, 1982, Comparison of the Upper Silurian *Atrypoidea* faunas of Arctic Canada and southern China: Journal of Paleontology, v. 56, p. 924–937.

Langenstrassen, F., 1972, Fazies und Stratigraphie der Eifel-Stufe im östlichen Sauerland: Göttinger Arbeiten zur Geologie und Paläontologie, no. 12, 106 p.

Liu Fa and Huang Zhu-xi, 1977, The Silurian rocks and faunas from Central Jilin Province: Journal of the Changchun Geological Institute (Quarterly), no. 1, p. 52–67 (Jilin Sheng Zhongbu de Zhiliuji diceng ji dongwuquxi, Changchun Dizhi Xueyuan Xuebao, 1977, 1 qi, 52-67 ye).

Mansuy, H., 1908, Contribution a la Carte Géologique de L'Indo-chine (Paléontologie): Direction Generale des Travaux Publics, Service des Mines, 73 p.

Marr, J. E., and Nicholson, H. A., 1888, The Stockdale Shales: Quarterly Journal of the Geological Society, v. 44, p. 654–732.

Modzalevskaya, E. A., 1968, Biostratigraphic subdivision of the Devonian in the Far East and Transbaikal Region, *in* D. H. Oswald, ed., Proceedings of the International Symposium on the Devonian System, Calgary 1967: Alberta Society of Petroleum Geologists, v. 2, p. 543–556.

Mu En-zhi, Boucot, A. J., Chen Xu, and Rong Jia-yu, 1986, Correlation of the Silurian Rocks of China: Geological Society of America Special Paper 202, 80 p.

Nikiforova, O. I., and Sapelnikov, V. P., 1971, Novie Rannesiluriiskie Virgianidae (Brachiopoda): Paleontologicheskiy Zhurnal, no. 2, p. 47–56.

Nitecki, M. H., 1972, North American Receptaculitid Algae: Fieldiana-Geology, v. 28, 108 p.

Pan Jian, Wang Shi-tao, Gao Lian-da, and Hou Jing-peng, 1978, Devonian Continental Sedimentary Formation of South China: Peking, Geological Publishing House, Editorial Committee of Professional Papers of Stratigraphy and Palaeontology, Chinese Academy of Geological Sciences, Professional Papers of Stratigraphy and Palaeontology, p. 240–269 (Huanan

Luxiang Nipenxi, Huanan Nipenxi Lunwenji, 240-269 ye).

Patte, E., 1926, Études Paléontologiques relatives à la Géologie de l'Est du Tonkin (Paléozoïque et Trias): Bulletin du Service Géologique de L'Indochine, v. 15, pt. 1, 240 p.

Qin Feng and Gan Yi-yan, 1976, The Palaeozoic stratigraphy of western Qin Ling Range: Acta Geologica Sinica, no. 1, p. 74–97 (Xiqinling Gushengdai diceng. Dizhi Xuebao, 1976, 1 qi, 74-97 ye).

Rong, Jia-yu, 1979, The *Hirnantia* fauna of China with comments on the Ordovician-Silurian boundary: Acta Stratigraphica Sinica, v. 3, no. 1, p. 1–29 (Zhongguo de Henantebei dongwuqun [*Hirnantia* fauna] binglun Aotaoxi yu Zhiliuxi de fenjie. Dicengxue Zazhi, Juan 3, qi 1, 1-29 ye. tuban 1-2).

Rong Jia-yu and Yang Xue-chang, 1977, On the *Pleurodium* and its relative genera: Acta Palaeontologica Sinica, v. 16, no. 1, p. 73–80 (Lun Lefangbei [*Pleurodium*] ji qi xiangguan de shu. Gushengwu Xuebao, Juan 16, qi 1, 73-80 ye).

—— , 1978, Silurian spiriferoids from Southwest China with special reference to their stratigraphic significance: Acta Palaeontologica Sinica, v. 17, no. 4, p. 347–384 (Xinan diqu zhiliuxi de shiyan ji qi diceng yiyi. Gushengwu Xuebao, Juan 17, qi 4, 357-384 ye).

—— , 1980, Brachiopods from the Miaokao Formation (Upper Silurian) of Qujing, Eastern Yunnan: Acta Palaeontologica Sinica, v. 19, no. 4, p. 263–288 (Diandong Qujing Shang Zhiliu Tong Miaogao Zu Wanzu Huashiqun. Gushengwu Xuebao, Juan 19, qi 4, 263-288 ye).

—— , 1981, Middle and late Early Silurian brachiopod faunas in Southwest China: Nanjing, Science Press, Academia Sinica, Nanjing Institute of Geology and Palaeontology Memoir 13, p. 163–270 (Xian Diqu Zaozhiliushi zhongwan qi de wanzu dongwuqun. Zhonguo Kexueyuan Nanjing Dizhi Gushengwu Yanjiusuo jikan, di 13 hao. Kexue Chubanshe).

Rong Jia-yu and Zhang Zi-xing, 1982, A southward extension of the Silurian *Tuvaella* Fauna: Lethaia, v. 15, p. 133–147.

Rong Jia-yu, Xu Han-kui, and Yang Xue-chang, 1974, Silurian Brachiopods, *in* The Handbook of Stratigraphy and Palaeontology of Southwest China, edited by Nanjing Institute of Geology and Palaeontology, Academia Sinica: Nanjing, Science Press, p. 195–208 (Zhiliuji Wanzu dongwu, Zhongguo Kexueyuan Nanjing Dizhi Gushengwu yanjiu suo bainzhu: Xinan diqu diceng gushengwu shouce, ye 195-208).

Rong Jia-yu, Johnson, M., and Yang Xue-chang, 1984, Early Silurian (Llandovery) Sea-Level changes in the Upper Yangzi Region of Central and southwestern China: Acta Palaeontologica Sinica, v. 23, no. 6, p. 672–694 (Shangyangziqu Zaozhiliushi de Haipingnian Bianhua. Gushengwu Xuebao, Juan 23, qi 6, 672-694 ye).

Rong Jia-yu, Su Yang-zheng, and Li Wen-guo, 1985, Brachiopods of the Xibiehe Formation (Upper Silurian) in Darhan Mumingan Joint Banner, Inner Mongolia, *in* Li Wen-guo, Rong Jia-yu, and Dong De-yuan, eds., Silurian and Devonian rocks and faunas of the Bateaobao area in Darhan-Mumingan joint Banner, Inner Mongolia: The People's Publishing House of Inner Mongolia, p. 27–48 (in Chinese), p. 51–53 (English summary).

Ruan, Yi-ping, Wang Cheng-yuan, Wang Zhi-hao, Rong Jia-yu, Mu Dao-cheng, Kuang Guo-dun, Yin Bao-an, and Su Yi-bao, 1979, On the age of Nahkaoling and and Yukiang Formation: Acta Stratigraphica Sinica, v. 3, no. 3, p. 225–229 (Lun Nagaolingzu he Yujiangzu de shidai, Dicengxue zazhi, 3 Juan, 3 qi, 225-229 ye).

Rubel, M., 1970, Brakhiopodi Pentamerida i Spiriferida Silura Estonii: Tallinn, USSR, Akademiya Nauk Estonskoy SSR Izvestiya, Khimiya-Geologiya, Izdatel'stvo "Valgus," 75 p.

Savarese, M., Gray, L. M., and Brett, C. E., 1986, Faunal and lithologic cyclicity in the Centerfield Member (Middle Devonian; Hamilton Group) of Western New York; A reinterpretation of depositional history, *in* Brett, C. E., ed., Stratigraphy and Depositional Environments of the Middle Devonian Hamilton Group in New York State: New York State Museum and Science Service Bulletin 457, p. 32–56.

Sheehan, P. M., 1977, Late Ordovician and earliest Silurian Meristellid brachiopods in Scandinavia: Journal of Paleontology, v. 51, p. 23–43.

Struve, W., 1982, The great gap in the record of marine Middle Devonian: Courier Forschungs institut Senckenberg, v. 55, p. 433–448.

Su Yang-zheng, 1976, The brachiopods from Cambrian to Devonian, *in* Atlas of Fossils of Northwest China, edited by Geological Bureau of Nei Monggol and Shenyang Institute of Geology and Mineral Resources (Northeast China Geological Sciences): Peking, Geological Publishing House, Nei Monggol volume, pt. I, Palaeozoic, p. 159–227 (Hanwuji-Nipenji de Wanzu dongwu. Huabei Diqu Gushengwu tuce, Nei Monggol fence 1, 159-227 ye. Dizhi Chubanshe).

—— , 1981, On the geological and geographical distribution of *Tuvaella* with reference to its habitat: Acta Palaeontologica Sinica, v. 20, no. 6, p. 567–576 (Lun Tuwabei de shikong fenbu he shengtai huanjing. Gushengwu Xuebao, Juan 20, qi 6, 567-576 ye).

Vladimirskaya, E. V., 1972, O sistematicheskom Polozhenii i geologicheskom raspostranenii roda *Tuvaella* (Brachiopoda): Pal. Zhur., no. 1, p. 37–44.

Vladimirskaya, E. V., and Chekhovich, V. D., 1969, Biostratigrafiya silura Tuvi (po materialam opornogo razreza "Elegest"): Geol. i Geofiz., no. 4, p. 11–19.

Wan Zheng-quan, Chen Yuan-ren, and Xu Qing-jian, 1978, Phylum Brachiopoda, *in* Atlas of Fossils of Southwest China, Sichuan Volume; Geological Publishing House, pt. 1, p. 284–380 (Wanzu dongwu men, Xinan diqu gushengwu tuce, Sichuan fence [1], 284-380 ye).

Wang Cheng-yuan, Ruan Yi-ping, Mu Dao-cheng, Wang Zhi-hao, Rong Jia-yu, Yin Bao-an, Kuang Guo-dun, and Su-Yi bao, 1979, Subdivision and correlation of the lower and Middle Devonian series in difference facies of Guangxi: Acta Stratigraphica Sinica, v. 3, no. 4, p. 305–311 (Guangxi butong xiangqu xiazhong nipentong de huafen he duibi, Dicengxue zazhi, 3 Juan, 4 qi, 305-311 ye).

Wang Yu, 1955, New genera of Brachiopoda: Acta Palaeontologica Sinica, v. 3, no. 1, p. 82–114 (Wanzulei de xinshu. Gushengwu Xuebao, Juan 3, qi 1, 82-114 ye).

Wang Yu and Rong Jia-yu, 1979, Brachiopod fauna of the Yukiang Formation and its paleobiogeographical significance, *in* Abstracts of Papers, 12th Annual Conference and 3rd National Congress of the Palaeontological Society of China: Palaeontological Society of China, p. 80–81 (Yujiangzu de Wanzu dongwuqun jiqi zai gushengwu dili quxi shang de yiyi. Zongguo Gushengwu Xuehui di 12 jie xueshu nianhui ji 3 jie quanguo huiyuan daibiao dahui Xueshu Lunwen zhaiyaoji, 80-81 ye).

Wang Yu, Yu Chang-min, and Wu Qi, 1974, Advances in the Devonian biostratigraphy of South China: Nanjing, Science Press, Academia Sinica, Nanjing Institute of Geology and Palaeontology Memoir 6, p. 1–45.

Wang Yu and Yu Chang-min, 1962, The Devonian System of China: Peking, Science Press (Zhongguo de Nipen Xi. Kexue Chubanshe), p. 1–140.

Wang Yu and Zhu Rui-fang, 1979, Beiliuan (Middle Middle Devonian) Brachiopods from South Guizhou and central Guangxi: Palaeontologica Sinica, no. 158, new series B, no. 15, p. 1–95 (Qiannan Guizhong Zhongnipenshi Beiliuqi Wanzu Dongwu. Zhongguo Gushengwu zhi, 158 ce, Xinyizhong 15 hao. 1-95 ye).

Wang Yu, Yu Chang-min, Xu Han-kui, Liao Wei-hua, Cai Chong-yang, 1979, Devonian biostratigraphy of South China: Acta Stratigraphica Sinica, v. 3, no. 2, p. 81–89 (Huanan Nipenji Shengwu diceng. Dicengxue Zazhi, 3 Juan, 2 qi, 81-89 ye).

Wang Yu, Rong Jia-yu, and Yang Xue-chang, 1980, The genus *Atrypoidea* (Brachiopoda) of Southwest China and its stratigraphical significance: Acta Palaeontologica Sinica, v. 19, no. 2, p. 100–116 (Zhongguo Xinan diqu de Fangwudongbei [*Atrypoidea*] ji qi diceng yiyi. Gushengwu Xuebao, Juan 19, qi 2, 100-116 ye).

Wang Yu, Boucot, A. J., Rong Jia-yu, and Yang Xue-chang, 1984, Silurian and Devonian biogeography of China: Geological Society of America Bulletin, v. 95, p. 265–279.

Xian Si-Yuan and Jiang Zong-long, 1978, Phylum Brachiopoda Dumeril, 1806, *in* Atlas of Fossils of Southwest China, edited by Working Team for Stratigraphy and Palaeontology of Guizhou: Peking, Geological Publishing House, Guizhou Volume, pt. I, from Sinian to Devonian, p. 251–337 (Wanzu Dongwu Men [Brachiopoda Dumeril, 1806]. Xinan Diqu

Gushengwu tuce, Guizhou fence [1], 251-337 ye. Dizhi chubanshe).

Xu Han-kui, 1977, Early Middle Devonian Plicanopliids from Nandan of Guangxi: Acta Palaeontologica Sinica, v. 16, no. 1, p. 59–70 (Guangxi Nandan zhong Nipenshi zaoqi de zhewuzhoubei lei [plicanopliids]. Gushengwu Xuebao, 16 Juan, 1 qi, 59-70 ye).

——, 1979, Brachiopods from the Tangxiang Formation (Devonian) in Nandan of Guangxi: Acta Palaeontologica Sinica, v. 18, no. 4, p. 362–380 (Guangxi Nandan Xian Nipenxi Tangxiangzu de Wanzulei. Gushengwu Xuebao, 18 Juan, 4 qi, 362-380 ye).

Xue Chun-ting, Su Yang-zheng, Zhang Hai-ri, and Cui Ge, 1980, Upper Silurian and Lower Devonian of the northwestern Xiao Hinggan Ling (Lesser Khingan Mountains): Journal of Stratigraphy, v. 4, no. 1, p. 1–12 (Xiao Xingan Ling Xibeibu Wanzhiliushi ji Zaonipenshi diceng. Dicengxue Zazhi, Juan 4, qi 1, 1-12 ye).

Yang Xue-chang and Rong Jia-yu, 1982, Brachiopods of the Upper Xiushan Formation in Sichuan, Guizhou, Hunan, and Hubei Province boundary regions: Acta Palaeontologica Sinica, v. 21, no. 4, p. 417–434 (Chuan, Qian, Xiang, E bianqu Xiushan zu Shangduan de Wanzu huashi).

Yin Zan-xun (Yin, T. H.), 1938, Devonian fauna of the Pochiao Shale of eastern Yunnan: Geological Society of China Bulletin, v. 18, p. 33–66.

Yu Chang-min and Kuang Guo-dun, 1986, Biostratigraphy, biogeography, and paleoecology of Devonian rugose corals from the Beiliu Formation in Beiliu, Guangxi: Acta Palaeontologica Sinica, v. 25.

Yu Chang-min and Yin Bao-an, 1978, A new stratigraphic unit of Lower Devonian in central Guangxi: Acta Stratigraphica Sinica, v. 2, no. 1, p. 23–31 (Guangxi zhongbu Nipenxi yige xinde diceng danwei-Ertang Zu. Dicengxue Zazhi, 2 Juan, 1 qi, 23-31 ye).

Yue Sen-xun and Bai Shun-liang, 1978, Devonian rocks of Dale area, Xiangzhou, Guangxi, *in* The Symposium on Devonian of South China, edited by Institute of Geological and Mineralogical Research, Chinese Academy of Geological Science: Peking, Geological Publishing House, p. 43–62 (Guangxi Xiangzhou Dale diqu Nipenji diceng).

Zhang Wen-tang, Chen Xu, Xu Han-kui, Wang Jun-geng, Lin Yao-kun, and Chen Jun-yuan, 1964, Silurian of northern Guizhou, *in* Paleozoic Rocks of Northern Guizhou: Nanjing Institute of Geology and Palaeontology, Academic Sinica, p. 79–110 (Guizhou beibu de Zhiliuxi, Guizhou beibu de gushengdai diceng, 79-110 ye, Zhongguo Kexueyuan Nanjing dizhi gushengwu yanjiu suo).

Zhang Zi-xin, Rong Jia-yu, and Di Qiao-ling, 1986, Silurian *Tuvaella gigantea* faunule (Brachiopoda) of the Barkol area, Northeastern Xinjiang: Acta Palaeontologica Sinica, v. 25 (Xinjiang dongbei bu Barkol diqu Zhiliuxi de daxing Tuwabei zuhe).

Typeset by WESType Publishing Services, Inc., Boulder, Colorado
Printed in U.S.A. by Malloy Lithographing, Inc., Ann Arbor, Michigan